Perfect Form

Perfect Form

VARIATIONAL PRINCIPLES, METHODS, AND
APPLICATIONS IN ELEMENTARY PHYSICS

Don S. Lemons

PRINCETON UNIVERSITY PRESS

PRINCETON. NEW JERSEY

Copyright © 1997 by Princeton University Press
Published by Princeton University Press, 41 William Street,
Princeton, New Jersey 08540
In the United Kingdom: Princeton University Press, Chichester,
West Sussex
All Rights Reserved

Library of Congress Cataloging-in-Publication Data
Lemons, Don S. (Don Stephen), 1949–
Perfect Form: variational principles, methods, and applications
in elementary physics / Don S. Lemons.
p.   cm.
Includes bibliographical references and index.
ISBN 0-691-02664-5 (alk. paper). —ISBN 0-691-02663-7
(pbk. : alk. paper)
1. Calculus of variations. 2. Mathematical physics. I. Title.
QC20.7.C3L46   1997
530.1'5564—dc20   96-9639

This book has been composed in Times Roman

Princeton University Press books are printed on
acid-free paper, and meet the guidelines for
permanence and durability of the Committee
on Production Guidelines for Book Longevity
of the Council on Library Resources

Printed in the United States of America
by Princeton Academic Press

10 9 8 7 6 5 4 3 2 1

10 9 8 7 6 5 4 3 2 1
(Pbk.)

ISBN-13: 978-0-691-02663-3
ISBN-10: 0-691-02663-7

Contents

Preface

> One can say that he who acts perfectly is . . . like
> a talented author who encloses a maximum of
> realities in the least possible volume.
> —Gottfried Wilhelm Leibniz, *Discourse on*
> *Metaphysics,* 1686

WHAT DOES THE PATH taken by a ray of light share with the curve of a wheat stalk bending in the breeze, or the trajectory of a thrown baseball? Each shape is different and each is a subject of different study. Yet all are optimal shapes. The path taken by a light ray is the quickest possible one. Light rays minimize travel time. Likewise, the bending stalk of wheat and the baseball trajectory minimize or maximize important physical quantities.

Almost all natural curves and shapes as well as many artificial ones are optimal with respect to some quantity. Natural curves are so because physical laws can be expressed as optimizing principles. Thus ray optics follows from Fermat's Principle of Least Time, and dynamics is governed not only by Newton's laws of motion but also and, equivalently, by the Principle of Least Action. Then, too, engineers seek the least expensive, the strongest, or the most efficient in their designs. Although the criteria may vary, in each the optimal is sought.

The attraction of perfect form is an old and powerful one. The Greeks, our teachers in all that is beautiful and precise, posed several mathematical problems whose solutions are either extreme values or extremizing curves. Hero's problem (ca. 125 B.C.E.) was to determine the angles formed by incident and reflected light at a mirror given the supposition that light "strives to move over the shortest possible distance, since it has not the time for slower motion." Just how a ray "strives," was then unclear. Does it send out forerunners or try out various paths before deciding on which to embark? Does the lure of perfection in any sense cause the ray's motion?

Hero's supposition is a statement of final cause. The final cause (that light minimizes its travel time between two points) of an effect (the light ray's path) is that for which the effect is achieved. Final causes seem strange to us because they locate the cause of motion in results rather than in prior conditions. Yet, since the time of Aristotle (ca. 350 B.C.E.), natural laws have been expressed in terms of final cause. Leonard Euler (1707–83), possibly the most productive and versatile of all mathematicians, wrote that "there is absolutely no doubt that every effect in the universe can be explained as satisfactorily from final causes, by the aid of the method of maxima and minima, as it can from the effective causes."[1]

"Variational principles," such as Fermat's, are the contemporary descendants of final cause. Almost every fundamental and derivative law of physics can be derived from an appropriate variational principle. Furthermore, Euler's "method of maxima and minima," now called the "calculus of variations," has developed into a standard tool of applied mathematics and is the natural language of variational principles. In a certain sense, variational principles and methods achieve the long sought after theoretical framework in which a rich variety of consequences flow from simple hypotheses.

With this volume I seek to introduce and give an elementary account of variational principles and methods for students of physics. In doing so I rely on several guiding lights. First, this book builds upon but does not extend greatly beyond the physics and mathematics usually encountered in introductory course sequences. The reader will find ray optics, particle trajectories, and Euler-Lagrange equations but not stress tensors, nonholonomic constraints, or Hamilton-Jacobi equations. Neither is the text concerned with versions of Hamilton's principle that cannot be expressed in terms of a variation of a definite integral. The mathematics assumed extends only to solving ordinary differential equations with the technique of separation of variables.

Second, the topics included are motivated by and organized around physical theories and their applications. Mathematical concepts and procedures are developed, as they were originally by Newton, Leibniz, Euler, and others, as a language with which to articulate and solve physical problems.

Third, and finally, the book begins with the most plausible and most restricted variational principles, such as Fermat's and Least Potential Energy, and develops more powerful ones through generalization. A widely adopted alternative is to postulate Hamilton's Principle, or deduce it from D'Alembert's Principle, and from these deduce all other results. The approach taken here slowly unfolds the subject.

---

[1] Leonhard Euler, *Methodus Inveniendi Lineas Curvas* (Lausanne and Geneva, 1744), translated by W. A. Oldfather, C. A. Ellis, and D. M. Brown, pp. 76–77.

End-of-chapter problems are provided for the benefit of teachers and students and others who wish to test and deepen their understanding of the material. A few are meant to be mere examples; most cannot be solved by substitution into received formulae. In some cases problems illustrate for the first time concepts mentioned only briefly in the main text. Some problems introduce new kinds of applications while a few complete or even extend the theory. As a whole, they, along with the footnotes, constitute a subtext enriching the main text but not absolutely necessary to it. On the other hand, there are sections (1.4, 3.4, 4.6, 4.7, and 5.4) which can be skipped without loss of continuity if one needs to hurry through the material. There is even a selection of text (chapter 1, sections 4.1–4.4, 5.1, and 6.1) and problems which constitute a non-calculus-of-variations introduction to variational principles.

Typically, there is no single course in the undergraduate physics curriculum that brings together the contents of this book. While understandable, this is a pity. Variational principles and methods unify much of physics. Their study is fascinating and useful and also prepares one for more advanced work. If variational principles have not been sufficiently appreciated, then, possibly, the first exposure should be a more extensive one. I believe there is a need for this text as a supplement to existing courses, as a main text for special short courses devoted entirely to the subject, and as a guide for individual study.

If *Perfect Form* can meet this need, no small credit is due students at Amherst and Bethel Colleges, especially Ken Friesen, Matt Harms, Jeremy Pearce, Eric Peters, and Jonathan Zerger, who worked many of the problems and read much of the text. I am also grateful to my colleagues at Amherst and Bethel Colleges for creating an environment in which it was possible to write; to Galen Gisler of Los Alamos National Laboratory for reading and improving a portion of the text; to Marion Deckert for suggesting the title "Leibnizian Physics"; and to Willis Overholt for producing the figures. Finally, and with great pleasure, I dedicate *Perfect Form* to my wife and sons: Allison, Nathan, and Micah.

Perfect Form

Least Time

> Nature operates by the simplest and most
> expeditious ways and means.
> —Pierre De Fermat, *Analysis ad Refractiones*,
> 1662

THE PRINCIPLE OF LEAST TIME relates the length and orientation of a light ray to the time required for light to propagate along the ray path. All the facts of ray optics are a consequence of this principle.

1.1 IMPORTANT FACTS

Many of the important facts of ray optics were well known in Pierre de Fermat's lifetime (1601–65): (1) light travels in straight lines in uniform media, (2) light reflects from a mirror like a billiard ball bouncing from a pool table bumper, (3) when passing from a less dense material (e.g., air) into a more dense material (e.g., water) light rays incline (i.e., refract) toward the interface normal, and (4) light rays are reversible, that is, light can propagate in either direction along the same path. Facts number 2 and 3, having to do with reflection and refraction, were also understood quantitatively. According to the *law of reflection*, the angle the incident ray makes with the mirror normal, θ_i, equals the angle the reflected ray makes with the normal, θ_r, that is,

$$\theta_i = \theta_r. \tag{1.1.1}$$

(See fig. 1.1.) While, according to *Snell's Law*, the sines of the angles (θ_1 and θ_2) the refracted rays (incident and transmitted) make with the normal to the interface between two different media are proportional, that is,

$$\frac{\sin \theta_1}{\sin \theta_2} = n_{21} \tag{1.1.2}$$

(fig. 1.2) where n_{21} is the *relative index of refraction* characteristic of both

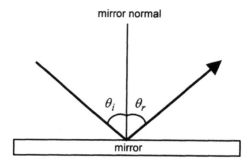

Fig. 1.1. The angle of incidence equals the angle of reflection

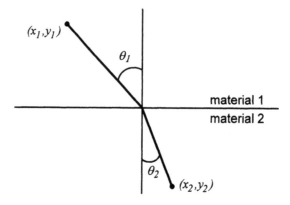

Fig. 1.2. Snell's Law: $\sin \theta_1 / \sin \theta_2 = \text{constant}$

media on either side of the interface. For both reflection and refraction, the incoming and outgoing rays and the surface normal lie in a single plane.

Just as important for this story is what Pierre de Fermat did not know: whether light traveled indefinitely fast or with finite speed. Not until 1675, did Olaf Roemer (1644–1710) infer a large but finite value for the speed of light in the vacuum of interplanetary space from the annual variation in the period of Jupiter's moon Io. Much less did the natural philosophers of the mid 1600s know the speed of light in the various media in which Snell's Law could be demonstrated.

1.2 AN INTERPRETATION

In 1662 Fermat boldly proposed a principle capable of correlating all the known facts of ray optics given the hypothesis that light travels more slowly in dense materials (e.g., glass or water) than it does in less dense materials (e.g., air). According to his *Principle of Least Time*, light propagates between two points in such a way as to minimize its travel time.

Snell's Law is easily derived from Fermat's Principle of Least Time. To do so, we search for a ray lying in the x–y plane which connects the two points (x_1, y_1) and (x_2, y_2) and minimizes the travel time. Again, refer to figure 1.2. Suppose light travels in a straight line through material 1 with speed v_1 from the point (x_1, y_1) to $(x, 0)$. Then it enters material 2 where the light speed is v_2 and goes on to (x_2, y_2). Given the ray's beginning (x_1, y_1) and ending (x_2, y_2) points, the single parameter x defines its path. The travel time $T(x)$ between the two points is

$$T(x) = \frac{\sqrt{(x - x_1)^2 + y_1^2}}{v_1} + \frac{\sqrt{(x_2 - x)^2 + y_2^2}}{v_2}. \tag{1.2.1}$$

The particular value of x which minimizes the travel time is one for which the derivative of the function $T(x)$ vanishes, that is,[1] $dT/dx = T'(x) = 0$ or

$$\frac{(x - x_1)}{v_1\sqrt{(x - x_1)^2 + y_1^2}} = \frac{(x_2 - x)}{v_2\sqrt{(x_2 - x)^2 + y_2^2}}. \tag{1.2.2}$$

Since by definition

$$\sin(\theta_1) = \frac{(x - x_1)}{\sqrt{(x - x_1)^2 + y_1^2}}, \tag{1.2.3}$$

et cetera, the condition (1.2.2) is equivalent to

$$\frac{\sin(\theta_1)}{v_1} = \frac{\sin(\theta_2)}{v_2}. \tag{1.2.4}$$

---

[1] Here and elsewhere, when there is only one independent variable, a prime will denote differentiation with respect to that variable; that is, $y' = y'(x) = dy/dx$ and $y'' = y''(x) = d^2y/dx^2$.

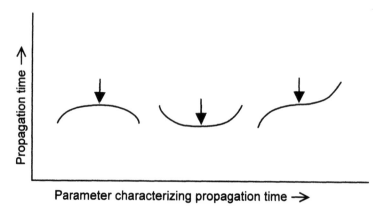

Parameter characterizing propagation time →

Fig. 1.3. Different kinds of stationary values of the propagation time function: (a) maximum, (b) minimum, and (c) inflection point

This derivation of Snell's Law also generates an expression for the relative index of refraction,

$$n_{21} = \frac{v_1}{v_2},$$ (1.2.5)

in terms of v_1 and v_2. When the first medium is a vacuum $v_1 = c$ and equation (1.2.5) defines an *absolute index of refraction*,

$$n_\alpha = \frac{c}{v_\alpha}.$$ (1.2.6)

Therefore, $n_{\alpha\beta} = n_\alpha/n_\beta$. In this notation Snell's Law becomes

$$n_1 \sin(\theta_1) = n_2 \sin(\theta_2).$$ (1.2.7)

The rule of reflection follows from the Principle of Least Time in a similar way. See problem 1.1 *Hero's Problem* at the end of the chapter. That the light ray lies in a plane formed by the two ray endpoints and the mirror normal also follows from the Principle of Least Time (problem 1.2 *More Than One Independent Variable*). Snell's Law can also be applied in a medium in which the index of refraction various continuously. See problem 1.3 *Continuum Limit*.

The condition $T'(x) = 0$ obtains at a minimum of the propagation time, but does not, by itself, guarantee a minimum. Other possibilities consistent with $T'(x) = 0$ (a maximum and an inflexion in $T(x)$) are illustrated in figure 1.3. If $T''(x) > 0$ at the x for which $T'(x) = 0$, then the path actually minimizes the travel time. For example, taking two derivatives of the travel time function (1.2.1), we get

$$\frac{d^2T}{dx^2} = \frac{y_1^2}{v_1((x-x_1)^2 + y_1^2)^{3/2}} + \frac{y_2^2}{v_2((x_2-x)^2 + y_2^2)^{3/2}}, \quad (1.2.8)$$

whose right-hand side is positive for all values of x. Therefore, Snell's Law corresponds to an actual minimum rather than a maximum or inflection of the travel time function. Indeed, since the graph of $T(x)$ is concave upward for all real x, the value of $T(x)$ for which $T'(x) = 0$ corresponds to an absolute and not only a local minimum.

In this and similar examples we follow a definite procedure. First, we formally describe and sometimes illustrate with a diagram a one-parameter family of geometrically possible rays. Each ray in the family must join the same two endpoints and each is distinguished by a unique parameter x. Second, we form the propagation time function $T(x)$. Third, we set $T'(x) = 0$ and solve for x. This selects a ray or rays, customarily called *true rays*, from among a family of possible ones. In principle, there is a fourth step: examining the sign of $T''(x)$ or, if necessary, higher derivatives of $T(x)$ for each true ray in order to distinguish between solutions which minimize the propagation time and those corresponding to maxima or inflections. We'll find in section 1.3 that the last step is unnecessary in ray optics. The procedure outlined here is straightforwardly generalized to accommodate situations in which two or more parameters are required to specify the family of geometrically possible rays. Problem 1.2 illustrates this generalization.

Fermat's hypothesis that light travels more slowly in more dense materials, and indirectly the Principle of Least Time itself, was verified for the first time in 1850 when Leon Foucault measured the speed of light in water and found it to be less than the speed in air by an amount predicted by the Principle of Least Time. In the meantime, Descarte (1596–1650), Newton (1642–1727), and Leibniz (1646–1716) had rival hypotheses which could account for refraction just as well as Fermat's.[2] Problem 1.4 *Least Resistance* formulates one of these ideas.

---

[2]Chapters 4 and 5 of Vasco Ronchi's *The Nature of Light* (Cambridge, Mass.: Harvard University Press, 1970) is a good general reference on the history of these ideas.

1.3 FERMAT'S PRINCIPLE

Descarte's followers were quick to point out a difficulty with the Principle of Least Time: the law of reflection, $\theta_i = \theta_r$, is sometimes consistent with the greatest rather than the least propagation time. Compare, for instance, two different ray paths both of which start at the center C of a spherical mirror of radius CA (fig. 1.4), reflect once, and return to a point B on the axis behind the center after reflection. One ray reflects at A and the other at D.

The ray path CDB which approaches the mirror head-on and returns along a diameter is the only one consistent with the reflection rule (1.1.1). It is also longer than neighboring paths such as CAB and for this reason has a longer propagation time. The path CDB is longer than the path CAB since twice CD equals CA + AE and path CB is greater than path EB. The latter is true because CB is opposite angle CEB and EB is opposite angle ECB and CEB > ECB. Note that angle ACE equals angle AEC by design. For an analytical proof that the ray CDB maximizes the propagation time, see problem 1.5 *Spherical Mirror*.

Therefore, rays obeying the law of reflection (1.1.1) occasionally maximize as well as minimize travel time. Apparently, nature is extravagant as well as economical. There are even valid ray paths which neither max-

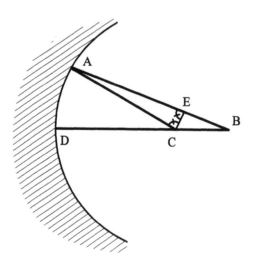

Fig. 1.4. Light sometimes follows the longer path (CDB) between two points (C and B)

imize nor minimize the travel time function. For instance, rays beginning at the center of a spherical mirror and reflecting from it anywhere on the surface and returning to the center all have the same path length and travel time. These and other exceptions to the Principle of Least Time involve reflection in special circumstances.

We meet these difficulties by stretching the content of the Principle of Least Time to include cases of greatest propagation time as well as other cases in which the propagation time function is merely a *stationary value* but not necessarily an extreme value. In a one-parameter family of rays, those parameters x which satisfy $T'(x) = 0$ are said to make $T(x)$ stationary.[3] An advantage of this formulation is that we need no longer worry about conditions on the second derivative of the time function; the vanishing of the first derivative alone is both necessary and sufficient for identifying a stationary value and thus a light ray's path. Properly speaking, this reformulated principle is a *Principle of Stationary Time*. However, more often than not, common usage attaches the old name to the new content or honors the principle's originator by replacing Principle of Least Time with the name *Fermat's Principle*. A similar reformulation of a "least principle" into a true *variational principle* also occurs in mechanics.

1.4 IMAGE FORMATION

Fermat's Principle solves all problems in geometrical optics. Here is one involving image formation. A luminous object at (x_1, y_1) emits rays in all directions. Some of these are intercepted by the lens (at $x = 0$) and focused to an image point (x_2, y_2). Figure 1.5 illustrates the geometry. What is the relation between image and object positions and properties of the lens?

Typically, the lens is a piece of glass whose thickness is a smooth function of the distance y from the lens symmetry axis at $y = 0$. However, a lens of uniform thickness could also be constructed with material of varying index of refraction. In either case the time $D(y)$ a ray takes to pass through the lens at position $(0, y)$ fully characterizes the lens. We are here concerned with a *thin lens* for which the delay D is not a function

---

[3]Precisely speaking, the stationary value of a function $f(x)$ is its value at a point x_o at which the function varies no more quickly than second order in small changes $\Delta x = x - x_o$ in x around x_o, that is, $f(x) \approx f''(x_o)\Delta x^2/2 + \cdots$.

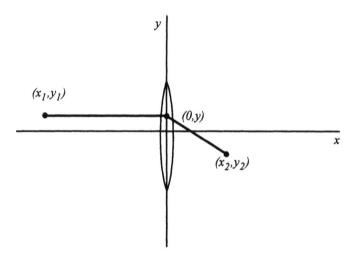

Fig. 1.5. Ray passing through a lens and connecting object and image points

of the angle with which a ray enters the lens. Let us also assume the lens is symmetric around the $y = 0$ axis so that $D(y) = D(-y)$.

The total propagation time T between the points (x_1, y_1) and (x_2, y_2) is given by

$$cT(y) = \sqrt{x_1^2 + (y_1 - y)^2} + \sqrt{x_2^2 + (y_2 - y)^2} + cD(y). \qquad (1.4.1)$$

When the *paraxial ray approximation* holds, that is, when $(y_{1,2} - y)^2 \ll x_{1,2}^2$, this expression reduces to

$$cT(y) = x_1 \left[1 + \frac{(y_1 - y)^2}{2x_1^2} \right] + x_2 \left[1 + \frac{(y_2 - y)^2}{2x_2^2} \right] + cD(y). \quad (1.4.2)$$

According to Fermat's Principle, the function $T(y)$ is stationary, that is, $T'(y) = 0$ or

$$-\frac{(y_1 - y)}{x_1} - \frac{(y_2 - y)}{x_2} + cD'(y) = 0 \qquad (1.4.3)$$

at the value of y at which a true ray, connecting points (x_1, y_1) and (x_2, y_2), passes through the lens.

That the lens *focus* true rays to an image point is a distinct and independent condition. For a perfectly sharp image all the true rays that come from the object point (x_1, y_1) and that are intercepted by the lens must be redirected by the latter to the image point (x_2, y_2). In other words, equation (1.4.3) must hold for all y. In practice, lenses are designed with delay properties that achieve a sharp image. A quadratic dependence,

$$D(y) = D(0) + D''(0)\frac{y^2}{2}, \tag{1.4.4}$$

will do the job if the lens curvature $D''(0)$ is chosen properly. Substituting (1.4.4) into condition (1.4.3) leads to

$$-\left(\frac{y_1}{x_1} + \frac{y_2}{x_2}\right) + y\left(\frac{1}{x_1} + \frac{1}{x_2} + cD''(0)\right) = 0. \tag{1.4.5}$$

Therefore, the requirements that rays be true and that the lens focus the true rays, regardless of y, to an image together produce the required conditions

$$\frac{y_1}{x_1} = -\frac{y_2}{x_2} \tag{1.4.6}$$

and

$$\frac{1}{x_1} + \frac{1}{x_2} = -cD''(0). \tag{1.4.7}$$

To the extent that $D(y)$ is not quadratic in y the focus is not perfectly sharp and so-called *spherical abberations* arise.

Since, according to (1.4.7), x_1 determines x_2 independently of y_1, image planes and not only image points exist. By definition, the *focal length* f of a lens is the image plane position, x_2, of objects at "infinity," that is, objects for which $x_1 \gg x_2$. According to this definition and equation (1.4.7),

$$f = -\frac{1}{cD''(0)}. \tag{1.4.8}$$

Therefore, the larger the lens curvature $D''(0)$, the smaller the focal length f. Equations (1.4.7) and (1.4.8) together imply

$$\frac{1}{x_1} + \frac{1}{x_2} = \frac{1}{f}. \tag{1.4.9}$$

Equations (1.4.6) and (1.4.7) solve the problem of finding the image position in terms of the object position and properties of the lens. They also recover familiar results usually derived by directly applying the law of refraction at both surfaces of the lens.[4] Variational methods easily

---

[4] See almost any introductory general physics book, for example, H. C. Ohanian, *Physics* (New York: Norton, 1989) pp. 926 ff.

generalize this approach to account for image formation with variously shaped lenses and mirrors.[5]

1.5 FINAL CAUSE

Fermat's Principle seems special among the laws of physics. But what accounts for its special character? Isn't it that in order to determine a ray path one must know the ray's destination in advance? The usual situation in mechanics is different: the pitcher's throw and the force exerted by the earth on a baseball is the cause of its parabolic trajectory and its arrival at first base. Mathematically, these two situations correspond to structurally different problems. A ray path is determined by the Principle of Least Time plus boundary values. The boundary values are the positions of the ray endpoints. In contrast, a typical problem in Newtonian dynamics is an initial value problem. The baseball trajectory is determined by Newton's Second Law plus the baseball's initial position and velocity.

In traditional language, Fermat's Principle plus boundary conditions is said to account for the phenomena of ray optics by postulating a *final cause*. A final cause operates when the result of a natural process (the ray endpoints and a minimum propagation time) determines the means by which that result is achieved (the true ray path). In contrast, our usual sense of causation is that of *efficient cause* in which a state of motion (the baseball's parabolic trajectory) is caused by the immediate prior condition (the throwing hand and the earth's pull).[6] The French polymath and inventor of the calculus, Gottfried Wilhelm Leibniz (1646–1716), was greatly attracted to explanation by final causes. He didn't deny that efficient causes are also useful, but for Leibniz "the way of final causes is easier, and is not infrequently of use for discerning important and useful truths which one would be a long time in finding by the other more physical route."[7]

Modern physics has largely, if not explicitly, adopted Leibniz's point of view. Variational principles such as Fermat's are mathematical ex-

---

[5] See D. S. Lemons, *Gaussian Thin Lens and Mirror Formulae From Fermat's Principle*, American Journal of Physics, vol. 62, pp. 367–68 (1994).

[6] For Aristotle a cause is that which is an answer to the question "Why?" His *Physics* distinguishes among four kinds of causes: material, formal, efficient, and final. See *The Basic Works of Aristotle*, Richard McKeon, ed. (New York: Random House, 1941); Stephanus, pp. 198a–198b; or text, pp. 247–48.

[7] Leibniz, *Discourse on Metaphysics* (Manchester: Manchester, 1953), p. 32.

pressions of final cause. Much, if not all, of modern physics can be written in terms of variational principles. They are useful not only because they are simply stated, have wide application, and solve problems easily, but also because their form serves as a template out of which new laws, such as those composing General Relativity, have been and can be generated.

Are there larger truths behind the usefulness and ubiquity of final cause? Some, in the rich history of our subject, would have it so.[8] Final cause suggests a purposive Intelligence guiding all things (Leibniz, Maupertuis, and Euler). Final cause is an expression of the simplicity and rationality of the world (Fermat and Leibniz). The Austrian physicist and philosopher of science Ernst Mach (1838–1916) was more skeptical. For him Fermat's principle, as well as all physical laws, at best, organize our observations and experimental results. "The principle does not so much promote our insight into true processes as it secures us a practical mastery of them. The value of the principle is of an economical character."[9]

CHAPTER 1 PROBLEMS

Problem 1.1 *Hero's Problem*
A ray begins above a flat reflecting surface at point (x_1, y_1), travels in a straight line to the surface at $(x, 0)$, reflects, and returns to a different point (x_2, y_2).

(a) Calculate the travel time function $T(x)$.

(b) Derive the rule of reflection (angle of incidence equals angle of reflection) from Fermat's Principle by solving $T'(x) = 0$.

(c) Show that the travel time function is an absolute minimum at the point for which the rule of reflection is satisfied by examining the sign of $T''(x)$. Hero of Alexandria (125 B.C.E.) was the first to solve a physics problem by invoking a minimum principle. According to Hero, light

---

[8]For historical and critical surveys, see, for example, Yourgrau and Mandelstam, *Variational Principles in Dynamics and Quantum Theory* (New York: Dover, 1968), chaps. 1 and 14, and Lanczos, *The Variational Principles of Mechanics* (New York: Dover, 1970), chap. X.

[9]Actually, this is Mach's evaluation of D'Alembert's Principle but it would seem also to apply to Fermat's Principle as well as all other physical principles. Ernst Mach, *The Science of Mechanics* (LaSalle, Illinois: Open Court, 1960), p. 430.

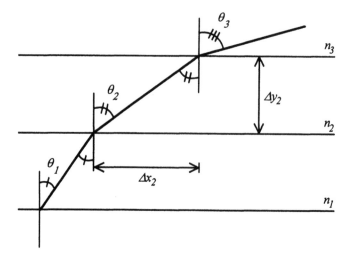

Fig. 1.6. Light propagation through a composite medium

"strives to move over the shortest possible distance, since it has not the time for slower motion."[10]

Problem 1.2 *More Than One Independent Variable*
In deriving Snell's Law (section 1.2) we assumed without proof that the incident and refracted rays and the surface normal all lie in the same plane. Show that this assumption follows from the Principle of Least Time by allowing the rays in figure 1.2 to begin at the point $(x_1, y_1, 0)$, end at the point $(x_2, y_2, 0)$, and meet the interface (the x–z plane) at the point $(x, 0, z)$.

(a) Express the propagation time function in terms of both x and z so that $T = T(x, z)$.

(b) Show that the necessary conditions for a minimum, that is, $\partial T/\partial x = 0$ and $\partial T/\partial z = 0$, lead to the desired results.

Problem 1.3 *Continuum Limit*
Light propagates through a medium composed of horizontal layers of equal thickness each with its own index of refraction as in figure 1.6. Show that in the continuum limit for which the index of refraction n is a function of y and $\Delta y/\Delta x \rightarrow dy/dx$ Snell's Law becomes $n(y)/\sqrt{1 + (dy/dx)^2} = $ constant.

---

[10]Quoted by W. Yourgrau and S. Mandelstam in *Variational Principles in Dynamics and Quantum Theory* (New York: Dover, 1979), p. 5.

Problem 1.4 *Least Resistance*
Consider the alternative theory that light is composed of particles which in passing from one medium to another minimize the so-called "resistance" $v_1 d_1 + v_2 d_2$. Here v_1, d_1 and v_2, d_2 are, respectively, the speeds and straight-line distances traveled in medium 1 and 2. The quantity $v_1 d_1 + v_2 d_2$ can be interpreted as a measure of the particle's rate of energy loss when its motion is resisted by a frictional force proportional to its speed v. Find a law of refraction analogous to Snell's Law implied by this *Principle of Least Resistance*. Show that, if $v_2 > v_1$, light will refract toward the normal upon entering medium 2. This behavior is contrary to that predicted by Snell's Law.

Problem 1.5 *Spherical Mirror*
Return to the geometry of figure 1.4. Let $R = CA = AE$, and $\theta =$ angle DCA. Find the propagation time function $T(\theta)$ and show that it is maximized for $\theta = 0$ and minimized for $\theta = \pi$.

Calculus of Variations

> All the greatest mathematicians have long since
> recognized that the method presented in this
> book is not only extremely useful in analysis, but
> that it also contributes greatly to the solution of
> physical problems.
> —Leonhard Euler, *Methodus Inveniendi*
> *Lineas Curvas,* 1744

FERMAT'S PRINCIPLE applies even when the set of possible light paths cannot be parameterized by a discrete set of variables. For this larger class of problems the ordinary techniques of the calculus do not suffice. Rather, one needs the *calculus of variations*, invented in part by Leonhard Euler (1707–83) in the early eighteenth century. The calculus of variations is the natural language of all variational principles.

2.1 AN INTRODUCTORY PROBLEM

We see straight shafts of light in a dusty room and readily affirm "light travels in straight lines." Can we derive this property of light from the Principle of Least Time? First, let's reduce the problem to a more familiar one by supposing that the light ray in a uniform medium chooses its path from among a one-parameter family of parabolas $y(x)$ lying in a single plane connecting two points on the x axis $(x_1, 0)$ and $(x_2, 0)$ so that

$$y(x) = \varepsilon(x - x_1)(x - x_2) \qquad (2.1.1)$$

where ε is the parabola's constant "curvature" and distinguishes among different members of the family (fig. 2.1). Note that the "straight parabola" has zero curvature ($\varepsilon = 0$). We determine the true ray by finding the parabola curvature ε which makes stationary the propagation time function $T(\varepsilon)$ by solving the equation $T'(\varepsilon) = 0$.

17

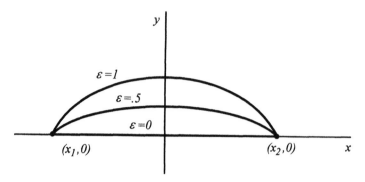

Fig. 2.1. Three paths between two points described by the curve
$y(x) = \varepsilon(x - x_1)(x - x_2)$

The integrated propagation time along a particular parabola is

$$T = \int \frac{ds}{v} \qquad (2.1.2)$$

or, equivalently,

$$T = \frac{1}{v} \int ds, \qquad (2.1.3)$$

since the speed of light v is constant throughout the presumed uniform
medium. Therefore, minimizing the propagation time between two points
is equivalent to minimizing the path length between two points. The differential arc length ds is related to changes dx and dy in the coordinates
x and y by the differential version of the Pythagorean theorem,

$$ds = \sqrt{dx^2 + dy^2} \qquad (2.1.4)$$

or

$$ds = \sqrt{1 + y'(x)^2}\, dx, \qquad (2.1.5)$$

so that (2.1.3) becomes the definite integral

$$T = \frac{1}{v} \int_{x_1}^{x_2} \sqrt{[1 + y'(x)^2]}\, dx. \qquad (2.1.6)$$

Differentiating equation (2.1.1) and substituting $y'(x)$ into the integrand
of (2.1.6) we arrive at

$$T(\varepsilon) = (1/v) \int_{x_1}^{x_2} \sqrt{1 + \varepsilon^2 (2x - x_1 - x_2)^2}\, dx, \qquad (2.1.7)$$

which describes how the propagation time depends upon the curvature ε of the parabola connecting the two points $(x_1, 0)$ and $(x_2, 0)$. Since the ray endpoints $(x_1, 0)$ and $(x_2, 0)$ are not functions of ε, we may differentiate (2.1.7) with respect to ε under the integral sign. Doing so, we find

$$T'(\varepsilon) = (1/v) \int_{x_1}^{x_2} (d/d\varepsilon)\sqrt{[1 + \varepsilon^2(2x - x_1 - x_2)^2]}\, dx \qquad (2.1.8)$$

or

$$T'(\varepsilon) = (1/v)\varepsilon \int_{x_1}^{x_2} \frac{(2x - x_1 - x_2)^2}{\sqrt{1 + \varepsilon^2(2x - x_1 - x_2)^2}}\, dx. \qquad (2.1.9)$$

Therefore, $T'(\varepsilon) = 0$ is solved by $\varepsilon = 0$. Since $\varepsilon = 0$ corresponds to the straight line $y = 0$, the light ray connecting two points in a medium with constant speed of light is straight. To solve this problem we used a *direct variational method*. Direct methods determine the parameters (e.g., ε) defining a particular functional form but not the form itself. They are powerful when we can intuit the latter but don't know the former. Direct methods become more accurate when higher order polynomials and more parameters are used to describe the form. In section 2.2, rather than merely refine the direct method, we will perfect it. But first let's review its generally useful features.

What we did was to use the derivative operator $(d/d\varepsilon)$ to compare the propagation time $T(\varepsilon)$ of various ray paths belonging to a one-parameter *comparison set* of rays. Then we selected that ray which extremized (actually, made stationary) $T(\varepsilon)$. In particular, the value of $T'(\varepsilon)$ at $\varepsilon = \varepsilon_o$ compares magnitudes of $T(\varepsilon)$ within a small neighborhood of the comparison set surrounding ε_o. For instance, if at $\varepsilon = \varepsilon_o$, $(d/d\varepsilon)T(\varepsilon) \neq 0$, then we know that neighboring values of $T(\varepsilon)$ are both larger and smaller than $T(\varepsilon_o)$; ε_o neither maximizes nor minimizes $T(\varepsilon)$. If, on the other hand, $T'(\varepsilon) = 0$ at $\varepsilon = \varepsilon_o$, ε_o is a stationary value of $T(\varepsilon)$; $T(\varepsilon)$ is neither larger nor smaller than $T(\varepsilon_o)$ through first order in ε within a small neighborhood of ε_o. According to Fermat's Principle, the stationary value ε_o identifies the true ray path from among the comparison set of possible rays.

The true ray selected with this method depends upon the quality and size of the comparison set. After all, the best from a barrel of rotten apples is not good. But, if we could select from all the apples in the world, we'd probably get one that was delicious. Likewise, the key to improving the variational method is to make the comparison set of possible rays as inclusive as possible. In doing so, we could increase the number of parameters which define the comparison set of ray paths from one (ε), to several $(\varepsilon_1, \varepsilon_2, \varepsilon_2)$, or even to a countably infinite number $(\varepsilon_1, \varepsilon_2, \varepsilon_3 \ldots)$. Then, the extremizing path is that determined by simultaneously solving the set of equations: $\partial T/\partial \varepsilon_1 = 0$, $\partial T/\partial \varepsilon_2 = 0, \ldots$. There is, however, a better way.

2.2 EULER-LAGRANGE EQUATION

Formally the problem is this: we seek a function $y(x)$ from among a maximally inclusive comparison set of continuous and twice differentiable but otherwise arbitrary functions $Y(x)$ connecting given endpoints, $[x_1, y(x_1)]$ and $[x_2, y(x_2)]$, that makes a particular definite integral,

$$I = \int_{x_1}^{x_2} f(x, Y, Y') \, dx, \qquad (2.2.1)$$

stationary. Here the integrand $f(x, Y, Y')$ is itself a continuous and twice differentiable function of x, Y, and Y'. Our notation underlines the distinction between the set of comparison functions $Y(x)$ and the particular member $y(x)$ which is actual or true. In ray optics $f(x, Y, Y')$ is chosen to render the integral I equal to the propagation time T.

Because the integral I is not a function of one or even a countably infinite number of discrete parameters, but is a function of a function, that is, a *functional* or expression which assigns a number to a function, a new method is required for finding the extremizing function $y(x)$. That new method is the calculus of variations. In this section we derive the simplest and most basic result of the calculus of variations; we solve the *first problem of the calculus of variations*.

Our derivation depends upon casting the new problem into the language of the old, discrete variable one. As before, we parameterize $Y(x)$ with ε and carefully choose ε so that $\varepsilon = 0$ reduces the comparison function $Y(x)$ to the true function $y(x)$. Thus by construction $I(\varepsilon)$ realizes a stationary value when $\varepsilon = 0$, that is, $I'(\varepsilon) = 0$ when $\varepsilon = 0$. We will exploit this property in determining the functional form of $y(x)$.

First, construct the comparison functions $Y(x)$ out of the supposed true or extremizing function $y(x)$ and another set of arbitrary functions $\eta(x)$ scaled by the parameter ε so that

$$Y(x) = y(x) + \varepsilon \eta(x). \qquad (2.2.2)$$

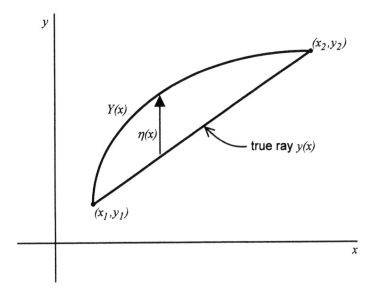

Fig. 2.2. True ray $y(x)$, a possible ray from the comparison set $Y(x)$, and the difference function $\eta(x)$

See figure 2.2. Since we limit the comparison set to continuous, twice differentiable functions and $y(x)$ is a special member of that set, the functions $\eta(x)$ are also continuous and twice differentiable. Then we can differentiate equation (2.2.2) and arrive at

$$Y'(x) = y'(x) + \varepsilon \eta'(x). \tag{2.2.3}$$

Furthermore, since the endpoints of all possible $Y(x)$ are the same, that is, $Y(x_1) = y(x_1) = y_1$ and $Y(x_2) = y(x_2) = y_2$,

$$\eta(x_1) = 0 \tag{2.2.4a}$$

and

$$\eta(x_2) = 0. \tag{2.2.4b}$$

Substituting expressions (2.2.2) and (2.2.3) for $Y(x)$ and $Y'(x)$ into the integral (2.2.1) yields

$$I(\varepsilon) = \int_{x_1}^{x_2} f(x, y(x) + \varepsilon \eta(x), y'(x) + \varepsilon \eta'(x)) \, dx. \tag{2.2.5}$$

This $I(\varepsilon)$ has the desired property: $\varepsilon = 0$ is a stationary value of $I(\varepsilon)$, that is, $I'(\varepsilon) = 0$ when $\varepsilon = 0$, because $\varepsilon = 0$ renders $Y(x) = y(x)$ which by construction makes the integral I stationary. Next, we differentiate $I(\varepsilon)$ with respect to ε under the integral sign of equation (2.2.5) and find that

$$I'(\varepsilon) = \int_{x_1}^{x_2} \left[\left(\frac{\partial f}{\partial Y} \right) \eta(x) + \left(\frac{\partial f}{\partial Y'} \right) \eta'(x) \right] dx \qquad (2.2.6)$$

where we have remembered that $Y = y + \varepsilon \eta$. Integrating the second term by parts yields

$$\begin{aligned} I'(\varepsilon) =\ & \int_{x_1}^{x_2} \eta(x) \left[\left(\frac{\partial f}{\partial Y} \right) - \frac{d}{dx} \left(\frac{\partial f}{\partial Y'} \right) \right] dx \\ & + \left(\frac{\partial f}{\partial Y'} \right) \eta(x) \bigg|_{x_2} - \left(\frac{\partial f}{\partial Y'} \right) \eta(x) \bigg|_{x_1}. \qquad (2.2.7) \end{aligned}$$

Since $\eta(x)$ vanishes at the endpoints, the two surface terms, $(\frac{\partial f}{\partial Y'})\eta(x)|_{x_2}$ and $(\frac{\partial f}{\partial Y'})\eta(x)|_{x_1}$, vanish and (2.2.7) reduces to

$$I'(\varepsilon) = \int_{x_1}^{x_2} \eta(x) \left[\left(\frac{\partial f}{\partial Y} \right) - \frac{d}{dx} \left(\frac{\partial f}{\partial Y'} \right) \right] dx. \qquad (2.2.8)$$

Finally, recall that $I(\varepsilon)$ was constructed so that $I'(\varepsilon) = 0$ when $\varepsilon = 0$ and also that $\varepsilon = 0$ collapses $Y(x)$ into $y(x)$. Therefore, setting $\varepsilon = 0$ changes equation (2.2.8) into

$$\int_{x_1}^{x_2} \eta(x) \left[\frac{\partial f}{\partial y} - \frac{d}{dx} \left(\frac{\partial f}{\partial y'} \right) \right] dx = 0. \qquad (2.2.9)$$

Now, the $\eta(x)$ are quite arbitrary except for continuity, smoothness, and vanishing endpoint conditions; otherwise, $\eta(x)$ may have many wiggles or none at all or it may vanish over part of its range and be very large in the rest. Integral (2.2.9) can vanish for each and every one of these diverse possibilities as required if and only if

$$\frac{\partial f}{\partial y} - \frac{d}{dx} \left(\frac{\partial f}{\partial y'} \right) = 0. \qquad (2.2.10)$$

If this were not the case and $(\partial f / \partial y) - (d/dx)(\partial f/\partial y') \neq 0$ for some values of x within a subinterval of the interval $x_1 \leq x \leq x_2$, then we could find a continuous, twice differentiable function $\eta(x)$ which vanished outside this subinterval but was positive (negative) wherever $(\partial f/\partial y) - (d/dx)(\partial f/\partial y') > (<)0$—thus contradicting the requirement

(2.2.9). This argument establishes the *fundamental lemma* of the calculus of variations.[1] Equation (2.2.10), known as the *Euler-Lagrange equation*, is the formal solution to the first problem of the calculus of variations. To summarize: the Euler-Lagrange equation (2.2.10) and the given boundary conditions $y(x_1) = y_1$ and $y(x_2) = y_2$ determine the function $y(x)$ which makes the definite integral $\int_{x_1}^{x_2} f(x, y, y')\, dx$ stationary.

The physical meaning of the integral (2.2.1) varies with context: the light propagation time, the potential energy, the action, or Hamilton's first principal function. However, in each application of a variational principle a global quantity expressed by an integral (2.2.1) reduces to a differential equation (2.2.10) relating differential changes among physical variables. A connection between the two kinds of statements, the global and the differential, is not hard to see. Consider, for instance, a true ray minimizing the light propagation time between two widely separated points in space. Then any segment of that true ray must also minimize the light propagation time between segment endpoints. If this were not the case, the ray segment could be changed to make the global propagation time even smaller, contrary to our initial supposition. Therefore, even differential sized seqments must minimize their light propagation time, and such requirement is naturally expressed in a differential equation.

The roots of the calculus of variations reach back into the late seventeenth century. The first seed was a problem posed by Johann Bernoulli (1667–1748) in 1696 as a challenge to "the shrewdest mathematicians of all the world."[2] (See problem 2.2 *Brachistochrone*.) Leonhard Euler first discovered the Euler-Lagrange equation (2.2.10). Joseph Lagrange (1736–1813) helped further develop the method. Today we recognize the calculus of variations as a generalization of the ordinary discrete-variable method of maxima and minima.

---

[1] See Weinstock, R., *Calculus of Variations* (New York: Dover, 1952), pp. 16–17.

[2] In fact, Johann meant to embarrass his older brother, Jacob (1654–1705), with whom he was publicly feuding at the time and whom he declared incompetent to solve the problem. The philosopher and mathematician Leibniz "solved the problem on the day he received Bernoulli's letter, and correctly predicted a total of only five solutions" within the alloted six-month time limit: they would come from the two Bernoullis, Newton, Leibniz, and L'Hospital. Newton, who was embroiled in a dispute with Leibniz over the priority of inventing the calculus, also solved the problem in one day and published his solution anonymously. On reading it, Johann, who was siding with Leibniz in that argument, recognized Newton's work, "just as from the pawprint, one recognizes the lion." See *Dictionary of Scientific Biography*, Charles Coulston Gillispie, ed. (New York: Scribner's, 1971), vol. I, p. 53.

2.3 FIRST INTEGRALS

In special cases the Euler-Lagrange equation integrates immediately to give one or more *first integrals*. A first integral of a differential equation is a function of the dependent and independent variables which contains derivatives of one order lower than the highest derivative in the differential equation and whose total derivative with respect to the independent variable vanishes. Two cases are important to us: (I) in which the integrand f does not depend on the dependent variable y, and (II) in which the integrand f does not depend on the independent variable x. In physical applications the absence of a dependence of either kind is said to signal the presence of a *symmetry*; conversely, each symmetry implies the existence of a first integral.

Case I: $f = f(x, y')$. Suppose the integrand f is invariant with respect to changes in y so that $\partial f / \partial y = 0$ or $f = f(x, y')$. Then, according to the Euler-Lagrange equation, (2.2.10),

$$\frac{d}{dx}\left(\frac{\partial f}{\partial y'}\right) = 0 \qquad (2.3.1)$$

so that $\partial f / \partial y'$ is a first integral, that is,

$$\frac{\partial f}{\partial y'} = C_1 \qquad (2.3.2)$$

where C_1 is an arbitrary constant whose value is determined by a boundary condition on $y(x)$. Note that while equation (2.3.1) is a second-order, ordinary, differential equation, (2.3.2) is a first-order one; thus the latter is a first integral of the former.

The propagation time integral of the introductory problem of section 2.1,

$$\int_{x_1}^{x_2} [1 + y'(x)^2]^{1/2} \, dx, \qquad (2.3.3)$$

results in this kind of first integral. Since the integrand is of the form $f(x, y')$, actually $f(y')$, the corresponding Euler-Lagrange equation integrates to the form (2.3.2), that is, to

$$\frac{y'(x)}{\sqrt{1 + y'(x)^2}} = C_1. \qquad (2.3.4)$$

Equation (2.3.4) is equivalent to the statement $y'(x) = \text{constant}$ which further integrated describes a straight line

$$y(x) = mx + b. \qquad (2.3.5)$$

The two constants m and b are determined by the two boundary conditions $y_1 = y(x_1)$ and $y_2 = y(x_2)$. Hence, a light ray connecting two points in a uniform medium is a straight line. See problem 2.3 *Minimum Surface* for another application of this first integral.

Case II: $f(y, y')$. Next, allow that the integrand f is not an explicit function of the independent integration variable x so that $\partial f/\partial x = 0$ or

$$f = f(y, y').$$
(2.3.6)

Differentiating (2.3.6) with respect to x and employing the "chain rule," this condition becomes

$$\frac{df}{dx} = \left(\frac{\partial f}{\partial y}\right) y' + \left(\frac{\partial f}{\partial y'}\right) y''.$$
(2.3.7)

Now, using the Euler-Lagrange equation (2.2.10) to eliminate the $(\partial f/\partial y)$ factor, we have

$$\frac{df}{dx} = y'' \left(\frac{\partial f}{\partial y'}\right) + y' \frac{d}{dx}\left(\frac{\partial f}{\partial y'}\right),$$
(2.3.8)

which is equivalent to

$$\frac{d}{dx}\left[f - y'\left(\frac{\partial f}{\partial y'}\right)\right] = 0.$$
(2.3.9)

Therefore,

$$f - y'\left(\frac{\partial f}{\partial y'}\right) = D$$
(2.3.10)

where D is an arbitrary constant. The Euler-Lagrange equation has again been reduced to a first-order ordinary differential equation.

2.4 MORE THAN ONE UNKNOWN FUNCTION

Fermat's Principle employs the propagation time integral while the latter uses the differential arc length or *line element ds*. The line element itself may be written in any set of coordinates and is said to characterize the coordinate-system *metric*. Commonly used coordinates (see fig. 2.3) and their line elements are: the Cartesian $x - y - z$ for which

$$ds = \sqrt{dx^2 + dy^2 + dz^2},$$
(2.4.1)

25

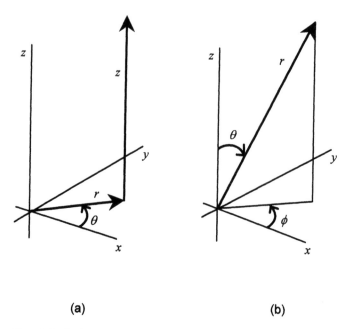

(a)   **(b)**

Fig. 2.3. Relation between Cartesian coordinates (x, y, z), (a) cylindrical polar coordinates (r, θ, z), and (b) spherical polar coordinates (r, θ, ϕ)

cylindrical polar $r - \theta - z$ for which

$$ds = \sqrt{dr^2 + (r d\theta)^2 + dz^2}, \qquad (2.4.2)$$

and spherical polar $r - \theta - \phi$ for which

$$ds = \sqrt{dr^2 + (r d\theta)^2 + (r \sin\theta d\phi)^2}. \qquad (2.4.3)$$

In order to speak generally about these and other coordinate systems we will adopt the following general notation for coordinates: $q_1, q_2, q_3 \cdots q_n$. Thus when $n = 3$ the set (q_1, q_2, q_3) could stand for (x, y, z), (r, θ, z), or (r, ϕ, θ) among other possibilities. These "q's" are traditional, if austere, symbols. We will use them when emphasizing, as we do below in generalizing sections 2.2 and 2.3, that a derivation or proof transcends specific features of a particular coordinate system.

When making an integral stationary, one of the coordinates, say q_1, is chosen to be the integration variable while the other $n - 1$ are the unknown functions, $q_2(q_1), q_3(q_1) \ldots q_n(q_1)$, to be determined. Thus q_1 is the independent and $(q_2, q_3, \ldots q_n)$ the dependent variables. When $n \geq 3$ there is more than one unknown function, and more than one Euler-Lagrange equation. Therefore, $n - 1$ comparison functions are defined by

$$Q_i(q_1) = q_i(q_1) + \varepsilon \eta_i(q_1) \qquad (2.4.4)$$

for $i = 2 \ldots n$ where, as before, the $\eta_i(q_1)$ are the true functions, the $\eta_i(q_1)$ are the arbitrary difference functions defining the comparison set, and ε is the comparison set parameter. As before, the comparison functions $Q_i(q_1)$ coincide at the integration endpoints q_{11} and q_{12}, that is, $\eta_i(q_{11}) = \eta_i(q_{12}) = 0$. Since the $q_i(q_1)$'s and the $\eta_i(q_1)$'s are continuous smooth functions, the derivative

$$Q_i'(q_1) = q_i'(q_1) + \varepsilon \eta_i'(q_1) \qquad (2.4.5)$$

exists. The integral to be made stationary,

$$I(\varepsilon) = \int_{q_{11}}^{q_{12}} f(Q_2, \ldots Q_n, Q_2', \ldots Q_n', q_1) \, dq_1, \qquad (2.4.6)$$

has, by construction, a vanishing derivative, $I'(\varepsilon)$, at $\varepsilon = 0$. Writing $I'(\varepsilon) = 0$ and setting $\varepsilon = 0$ we arrive at

$$\int_{q_{11}}^{q_{12}} \sum_{i=2}^{n} \left[\frac{\partial f}{\partial q_i} \eta_i(q_1) + \frac{\partial f}{\partial q_i'} \eta_i'(q_1) \right] dq_1 = 0. \qquad (2.4.7)$$

Integrating the second term in the brackets by parts, (2.4.7) becomes

$$\int_{q_{11}}^{q_{12}} \sum_{i=2}^{n} \left[\frac{\partial f}{\partial q_i} - \frac{d}{dt} \left(\frac{\partial f}{\partial q_i'} \right) \right] \eta_i(q_1) \, dq_1 = 0 \qquad (2.4.8)$$

which must obtain for arbitrary difference functions $\eta_i(q_1)$. In particular, (2.4.8) holds for the special case in which all the difference functions vanish identically except $\eta_k(q_1)$, that is, $\eta_i(q_1) = 0, i \neq k$. In this case, (2.4.8) reduces to

$$\int_{q_{11}}^{q_{12}} \left[\frac{\partial f}{\partial q_k} - \frac{d}{dt} \left(\frac{\partial f}{\partial q_k'} \right) \right] \eta_k(q_1) dq_1 = 0 \qquad (2.4.9)$$

which, given the fundamental lemma, implies

$$\frac{\partial f}{\partial q_k} - \frac{d}{dq_1} \left(\frac{\partial f}{\partial q_k'} \right) = 0. \qquad (2.4.10)$$

The latter follows because $\eta_k(q_1)$ remains arbitrary even though the other $\eta_i(q_1)$, $i \neq k$, have been specified. Since this argument may be repeated in turn for each of the $n - 1$ dependent variables q_k, the Euler-Lagrange equation (2.4.10) holds for each $k = 2 \ldots n$. Therefore, there is a separate Euler-Lagrange equation for each function $q_k(q_1)$. Note that the integration variable q_1 is the independent variable of each of these Euler-Lagrange equations.

First integrals of the $n - 1$ Euler-Lagrange equations (2.4.10) are simple extensions of those in the one dependent variable case. In particular, if the integrand f is independent of one of the dependent coordinates q_k ($k \neq 1$), that is, if $\partial f / \partial q_k = 0$, then the Euler-Lagrange equation for q_k reduces to the first integral

$$\frac{\partial f}{\partial q_k'} = C_k \text{ (a constant).} \tag{2.4.11}$$

If $\partial f / \partial q_k = 0$, q_k is called an *ignorable coordinate*. When the integrand f does not depend explicitly on the integration variable q_1, that is, $\partial f / \partial q_1 = 0$, we may, following the procedure of case II, section 2.2 (see problem 2.4 *First Integral*), show that

$$\sum_{i=2}^{i=n} q_i' \frac{\partial f}{\partial q_i'} - f = D \text{ (a constant).} \tag{2.4.12}$$

Equation (2.4.12) is, in essence, a first integral of the whole set ($k = 2, \ldots n$) of Euler-Lagrange equations (2.4.10).

CHAPTER 2 PROBLEMS

Problem 2.1 *Euler-Lagrange Equation*
Find the function $y(x)$, having boundary conditions $y(0) = 0$ and $y(1) = 1$, that makes the integral $\int_{x_1}^{x_2} (y'^2 + yy' + y^2) \, dx$ stationary.
[Answer: $y(x) = \frac{\sinh(x)}{\sinh(1)}$.]

Problem 2.2 *Brachistochrone*[3]
A smooth curved planar wire joins two points as shown in figure 2.4. A bead on the wire slides without friction from rest at the upper to the

---

[3] This problem and the next, *Minimum Surface*, have appeared in textbooks for scores of years if not for centuries.

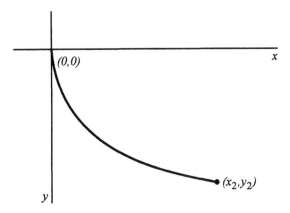

Fig. 2.4. A plane curve along which a massive bead
slides

lower endpoint under the influence of gravity. Its mechanical energy is
conserved as it moves along the wire. Choose down to be the positive y
direction.

(a) Show that the time T required for the bead to complete its journey
is given by

$$T = [1/\sqrt{(2g)}] \int_0^{x_2} \sqrt{\frac{(1 + y'^2)}{y}} \, dx.$$

(b) Given that the integrand of the above integral is independent of x,
show that the curve $y(x)$ making T stationary satisfies the differential
equation

$$dy/dx = [(b - y)/y]^{1/2}$$

where b is a constant.

(c) Change the dependent variable from y to ϕ where $y = b \sin^2 \phi/2$
and show that the above can be integrated to yield

$$x = (b/2)(\phi - \sin \phi).$$

Make use of common trigonometric identities.

This solution describes a *cycloid*, that is, the curve mapped out by a
point on the rim of a wheel as it rolls on a flat surface. In the present
context the cycloid is called a *brachistochrone* after the two Greek words
"brakhus" indicating "shortness" and "chronos" meaning "time."

Problem 2.3 *Minimum Surface*

A curve $y(x)$ in the $x-y$ plane connecting the points (x_1, y_1) and (x_2, y_2) is revolved around the y-axis and generates a surface of revolution with area

$$A = \int_{x_1}^{x_2} 2\pi x \sqrt{1 + y'(x)^2}\, dx.$$

Show that the curve $y(x)$ which generates the surface of revolution with the least area can be expressed in terms of the inverse hyperbolic cosine

$$y(x) = C_1 \cosh^{-1}(x/C_1) + C_2$$

where C_1 and C_2 are constants. In principle, C_1 and C_2 could be evaluated in terms of the endpoints (x_1, y_1) and (x_2, y_2).

Problem 2.4 *First Integral*

Show that

$$\sum_{i=2}^{i=n} q_i' \frac{\partial f}{\partial q_i'} - f = D \text{ (a constant)}$$

is a first integral by taking its total derivative with respect to q_1 given that the set of Euler-Lagrange equations,

$$\frac{\partial f}{\partial q_k} - \frac{d}{dq_1}\left(\frac{\partial f}{\partial q_k'}\right) = 0,$$

with $k = 2, \ldots n$ and also that the integrand f does not depend explicitly upon the integration variable q_1, that is, $\partial f/\partial q_1 = 0$.

Problem 2.5 *Variational versus Direct Method*

(a) Find the function $y(x)$ which connects the points $y(0) = 0$ and $y(1) = 1$ and makes the integral

$$I = \int_0^1 \left[y'^2 - \frac{\pi^2 y^2}{4}\right] dx$$

stationary by solving the Euler-Lagrange equation.
[Answer: $y(x) = \sin(\pi x/2)$.]

(b) Since the integrand f of the integral I is not a function of the independent variable x, $f - y' \partial f/\partial y'$ is a first integral of the Euler-Lagrange equation. Show that this first integral leads to the same extremizing function $y(x) = \sin(\pi x/2)$.

30

(c) A quadratic function $y(x; a)$ which connects the points $y(0; a) = 0$ and $y(1; a) = 1$ has the form

$$y(x; a) = (1 - a)x + ax^2$$

where a is its characterizing parameter. Find the particular value of the parameter a which makes stationary the integral I given that $y(x; a)$ defines the comparison set. Warning: Extensive but straightforward algebra is required.

[Answer: $a = -.819$]

(d) Compare values of the two extremizing functions $y(x)$ and $y(x; a)$ at $x = .5$ by calculating the fractional difference

$$\text{abs } \{[y(.5) - y(.5; a)]/y(.5)\}$$

for the value of a found in part (c).

[Answer: .00326]

Curved Light

> It is thus that we shall in the first place explain
> the refractions which occur in the air, which
> extend[s] from here to the clouds and beyond.
> The effects of which refractions are very
> remarkable; for by them we often see objects
> which the rotundity of the Earth ought otherwise
> to hide.
> —Christian Huygens, *Treatise On Light*, 1678

APPLYING Fermat's Principle when the index of refraction is a continuous function of spatial coordinates requires the calculus of variations. Several examples follow.

3.1 PLANAR ATMOSPHERE

We might reasonably suppose that the atmospheric index of refraction over scales for which the earth is a flat surface $y = 0$ is a function $n(y)$ of height y only. In such case the propagation time T along a ray connecting two endpoints (x_1, y_1) and (x_2, y_2) in a plane extending above and normal to the surface is given by

$$cT = \int_{x_1}^{x_2} dx\, n(y)\sqrt{1 + y'^2}. \tag{3.1.1}$$

The product of the vacuum speed of light c and the propagation time T is known as the *optical path length*. According to Fermat's Principle, the optical path length is stationary for a true ray path. Since the integrand

$$f(y, y') = n(y)\sqrt{1 + y'^2} \tag{3.1.2}$$

of the propagation time integral does not depend explicitly on the independent variable x, the resulting Euler-Lagrange equation reduces to the first integral $f - y'(\partial f/\partial y') = D$ (a constant) which in this case is

$$\frac{n}{\sqrt{1 + y'^2}} = D. \tag{3.1.3}$$

Evidently, the constant D is the value of the index of refraction at the point where the ray becomes horizontal, that is, where $y'(x) = 0$. Because the angle θ between the ray tangent and a vertical line (see fig. 3.1) is related to y' by $\sin \theta = 1/\sqrt{1 + y'^2}$, equation (3.1.3) can also be written as

$$n(y) \sin \theta = D. \tag{3.1.4}$$

Threfore, equation (3.1.3) or, equivalently, (3.1.4) generalizes Snell's Law to a planar atmosphere in which the index of refraction is a continuous function of the vertical coordinate y. Equation (3.1.3) reduces to the quadrature integral

$$x - x_o = \pm \int_{y_o}^{y} \frac{dy}{\sqrt{\frac{n(y)^2}{D^2} - 1}}. \tag{3.1.5}$$

For a number of functional forms $n(y)$ the integral (3.1.5) can be completed analytically. See, for example, problem 3.1 *Atmospheric Light Trajectory*, in which the index of refraction decreases linearly with height so that $n(y) = n_o - \lambda y$ where $\lambda > 0$.

Normally, the index of refraction decreases with altitude y, so that $(dn/dy) < 0$. Then the ray described by (3.1.5) is concave downward. Evidently, light minimizes its propagation time by arching its path upward between ray endpoints. As a consequence, objects are not where they seem to be but are actually a little lower than the direction in which our eyes look to see them. In inversion layers the index of refraction increases with altitude, $(dn/dy) > 0$, and the ray is concave upward. Thus mirages are formed.[1]

---

[1]A number of other interesting and unusual phenomena arise from curved rays. Mirages formed along roads and walls, the sighting of ships and islands over the earth's curvature, "looming," displacement of astronomical objects, "the green ray," "fata morgana," and other visible consequences of curved light rays are discussed by M. Minnaert, *The Nature of Light and Color in the Open Air* (New York: Dover, 1954), chap. IV. An early description of these kinds of phenomena and an account of "How consequently some objects appear higher than they are" is given by Christian Huygens in his *Treatise On Light* (New York: Dover, 1962).

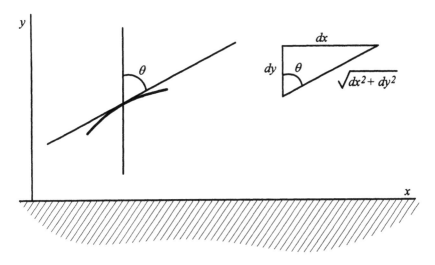

Fig. 3.1. In a planar atmosphere $n(y) \sin \theta = $ constant where $\sin \theta = 1/\sqrt{1 + y'^2}$

3.2 ROAD SURFACE MIRAGE

Sunshine warms a road and the adjacent air. An inversion layer forms in which the air lower down is hotter and less dense than that higher up. Throughout the layer the air pressure P is roughly constant because the weight of the column of air held up by the atmosphere changes only very slowly with altitude. In other words, the scale length over which the air pressure changes is much larger than the scale length over which the air temperature changes. Given that $P \propto \rho T$ and $P \approx$ constant, the air density ρ is smaller in the warm lower region than in the cool higher region. In short,

$$\rho T = \text{constant.} \tag{3.2.1}$$

Furthermore, in a gas the deviation of the index of refraction from its value in a vacuum ($n = 1$) is proportional to the gas density ρ, that is,

$$(n - 1) \propto \rho, \tag{3.2.2}$$

a relation known as the *Gladstone-Dale Law*. Therefore, in an inversion layer n is smaller lower down than higher up and light rays within it will be concave upward. These conditions are right for an inferior mirage.

Let's determine the distance L at which the rays forming a mirage appear to reflect from the road and see if it checks, at least roughly, with

35

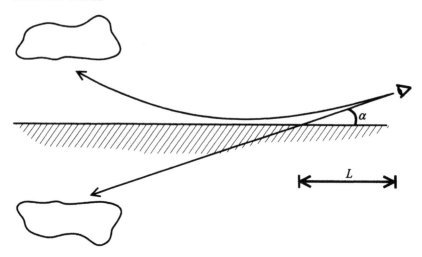

Fig. 3.2. Inferior mirage

our experience. For this pupose refer to figure 3.2. At the position in the hot air where the ray becomes horizontal, we denote the index of refraction by n_h; the subscript "h" is for hot. At eye position we denote the index of refraction by n_c ("c" is for cold) and the ray's angular deviation from the horizontal by α. The generalized Snell's Law (3.1.4) in this case is

$$n_c \sin(\pi/2 - \alpha) = n_h,\tag{3.2.3}$$

or

$$\cos \alpha = \frac{n_h}{n_c}.\tag{3.2.4}$$

Typically, $\alpha \ll 1$ so that $\cos \alpha \approx 1 - \alpha^2/2$ is a good approximation. Then the above is equivalent to

$$\alpha \approx \sqrt{2(1 - n_h/n_c)}.\tag{3.2.5}$$

Given the constant pressure condition (3.2.1) and the Gladstone-Dale law (3.2.2), we can write

$$(n_h - 1)T_h = (n_c - 1)T_c\tag{3.2.6}$$

from which we find that

$$1 - \frac{n_h}{n_c} = \left(1 - \frac{1}{n_c}\right)\left(1 - \frac{T_c}{T_h}\right).\tag{3.2.7}$$

Equation (3.2.5) for the angular deviation α then becomes

$$\alpha = \sqrt{2\left(1 - \frac{1}{n_c}\right)\left(1 - \frac{T_c}{T_h}\right)}. \qquad (3.2.8)$$

On a warm day the air temperature two meters from the ground might be $T_c \approx 303°$ (30°C or 86°F) and near the road surface $T_h \approx 323°$ (50°C or 122°F). The index of refraction of air at a given temperature can be found from a handbook of physical values. For $T_c = 303°$, $n_c = 1.00026$. Then $\alpha = 5.7 \times 10^{-3}$ radians ($\approx.32$ degree). Thus the eye at 2 meters above the road surface sees the mirage on the road at a distance $L = 2/(5.7 \times 10^{-3})$ or 351 meters—quite in keeping with everyday possibility. The larger the temperature variation, the closer the mirage appears.[2]

3.3 FIBER OPTIC

Fiber optics transmit light along their length by internal reflections from a discontinuity or gradient in their index of refraction. Imagine, then, a light ray in a cylindrically symmetric fiber optic with a nonuniform index of refraction $n = n(r)$. Appropriate coordinates are cylindrical coordinates r, θ, and z. According to Fermat's Principle, the ray path connecting two arbitrary points (r_1, θ_1, z_1) and (r_2, θ_2, z_2) in the fiber optic makes the optical path length

$$cT = \int n(r)\sqrt{dr^2 + (rd\theta)^2 + dz^2} \qquad (3.3.1)$$

stationary. Choosing z as the independent variable equation (3.3.1) becomes

$$cT = \int_{z_1}^{z_2} n(r)\sqrt{r'^2 + (r\theta')^2 + 1}\, dz \qquad (3.3.2)$$

where $r' = dr/dz$ and $\theta' = d\theta/dz$. The functions $r(z)$ and $\theta(z)$ describing a true ray are solutions to the two Euler-Lagrange equations

$$\frac{\partial f}{\partial r} - \frac{d}{dz}\left(\frac{\partial f}{\partial r'}\right) = 0 \qquad (3.3.3)$$

[2]There is, of course, nothing on the road; the image formed is virtual. It is only the tangent to the ray at the eye which intersects the road at distance L.

and

$$\left(\frac{\partial f}{\partial \theta}\right) - \frac{d}{dz}\left(\frac{\partial f}{\partial \theta'}\right) = 0, \tag{3.3.4}$$

respectively, where

$$f = n(r)\sqrt{r'^2 + (r\theta')^2 + 1} \tag{3.3.5}$$

is the integrand of (3.3.2). Since θ is an ignorable coordinate, that is, $\partial f/\partial \theta = 0$, the second of the two Euler-Lagrange equations (3.3.4) integrates immediately to

$$\frac{n(r)r^2\theta'}{\sqrt{r'^2 + r^2\theta'^2 + 1}} = C_1 \text{ (a constant).} \tag{3.3.6}$$

In addition, $f - r'\partial f/\partial r' - \theta'\partial f/\partial \theta' = D$ (a constant) since f is not an explicit function of the independent variable z. Thus we find that

$$\frac{n(r)}{\sqrt{r'^2 + r^2\theta'^2 + 1}} = D. \tag{3.3.7}$$

These two coupled first-order ordinary differential equations, (3.3.6) and (3.3.7), are sufficient for determining the two unknown functions $r(z)$ and $\theta(r)$ given the functional dependence $n(r)$. See problems 3.4 *Semicircular Rays* and 3.5 *Duffing's Equation* for special cases.

Equation (3.3.6) tells us that if upon entering the fiber optic $\theta'(0) = 0$, then the constant C_1 vanishes and $\theta' = 0$ (that is, $\theta = $ constant) for all z. Consequently, the ray remains within a single ($\theta = $ constant) plane containing the optic symmetry axis $r = 0$. Such rays are called *meridional rays*. If, on the other hand, the ray enters the fiber optic normal to the symmetry axis so that $r' \to \infty$ ($dz/dr \to 0$), then both C_1 and D vanish and $z = $ constant. Consequently, the ray is confined to a ($z = $ constant) plane normal to the optic axis $r = 0$. So-called *helical rays* confined to a cylindrical surface of constant radius r_o are also possible but only when the index $n(r)$ has a special profile. See problem 3.6 *Helix*. When none of these special conditions obtain, the rays are said to be *skew*.

3.4 PARAMETRIC RAY EQUATIONS

Typically, each problem may be parameterized in several ways and from among these we are free to choose the most convenient. For instance, choosing one parameter as integration variable might make the others ignorable. Then the Euler-Lagrange equations immediately produce a first integral for each ignorable coordinate. For the planar atmosphere of section 3.1, $n = n(y)$; we choose y as the independent integration variable, and, consequently, x is an ignorable coordinate. Likewise, in a cylindrical fiber optic, $n = n(r)$; we choose z as the independent integration variable, and θ is an ignorable coordinate.

Alternatively, the index of refraction may have no particular symmetry in any of the common coordinate systems. In that case, there is no reason to single out one of the coordinates as integration variable. Rather, we may choose yet another parameter as integration variable. It is only required that we conceive of each of the coordinates q_i as depending on the chosen integration parameter and that the integration is between definite limits. To illustrate, let's invent an integration parameter Ω. By definition, $\Omega = 0$ at the beginning of the ray, $\Omega = 1$ at the end of the ray, and the ray path changes continuously from beginning to end as Ω varies from 0 to 1.

In three-dimensional Cartesian x–y–z space with Ω as independent integration parameter, the integral to be made stationary is

$$cT = \int_0^1 n(x, y, z)\sqrt{x'^2 + y'^2 + z'^2}\, d\Omega \qquad (3.4.1)$$

where $x' = dx/d\Omega$, etc. Since there are three functions $x(\Omega)$, $y(\Omega)$, and $z(\Omega)$ and each one of these can be varied independently, there are three Euler-Lagrange equations of which the one determining $x(\Omega)$ is

$$\sqrt{x'^2 + y'^2 + z'^2}\frac{\partial n}{\partial x} = \frac{d}{d\Omega}\left(\frac{nx'}{\sqrt{x'^2 + y'^2 + z'^2}}\right). \qquad (3.4.2)$$

All three Euler-Lagrange equations may be expressed compactly in vector form as

$$\sqrt{x'^2 + y'^2 + z'^2}\nabla n = \frac{d}{d\Omega}\left(\frac{n\mathbf{x}'}{\sqrt{x'^2 + y'^2 + z'^2}}\right) \qquad (3.4.3)$$

where $\mathbf{x} = x(\Omega)\mathbf{e}_x + y(\Omega)\mathbf{e}_y + z(\Omega)\mathbf{e}_z$. Equations of this form are called *parametric ray equations*. The equations (3.4.3) are general laws of geometrical optics derived from Fermat's Principle with the help of the calculus of variations. Their structure reminds us of Newton's second law of motion, $\mathbf{F} = d\mathbf{p}/dt$, with the ∇n term playing the role of force, the parameter Ω the role of time, and $x'n/\sqrt{x'^2 + y'^2 + z'^2}$ the role of linear momentum.[3] We can always reparameterize an integration in this way although to do so is not always helpful. Once we have, we may without ambiguity rescale the independent parameter Ω and so identify it with either the commulative arc length s or propagation time t along the true ray. See, for instance, problem 3.9 *Time Evolution Equations*.

Suppose, for example, the index of refraction is that of a planar atmosphere for which $n = n(y)$. Now, if $z'(\Omega) = 0$ at both ray endpoints $\Omega = 0$ and $\Omega = 1$, then, according to the z component of the vector equation (3.4.3), $z = 0$ for all values of Ω and the ray is confined to the x–y plane. The x component of (3.4.3) integrates to

$$\frac{x'n(y)}{\sqrt{x'^2 + y'^2}} = C_1 \tag{3.4.4}$$

which along with the y component of (3.4.3) describes the true ray. Changing the independent variable in (3.4.4) from Ω to the spatial coordinate x recovers the local version of the generalized Snell's Law, that is, equation (3.1.3) or

$$\frac{n(y)}{\sqrt{1 + y'^2}} = C_1 \tag{3.4.5}$$

where here $y' = dy/dx$.

CHAPTER 3 PROBLEMS

Problem 3.1 *Atmospheric Light Trajectory*
Suppose an atmosphere has index of refraction $n(y) = n_o - \lambda y$.
  (a) Show that rays obey the differential equation

$$\frac{n_o - \lambda y}{\sqrt{1 + y'^2}} = D \text{ (a constant)}.$$

  (b) Solve this differential equation for initial values $y(0) = 0$ and $y'(0) = 0$. To do the resulting integral use the substitution $y \to \phi$ where $n_o - \lambda y = D \cosh\phi$.

[3]For an extended presentation of the ray optics-particle mechanics analogy, see the article " 'F = ma' Optics" by J. Evans and M. Rosenquist in the American Journal of Physics, vol. 54, pp. 876–83.

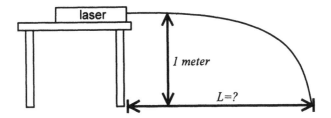

Fig. 3.3. Laser beam "falling" in a normal atmosphere

(c) Sketch your answer in the $x > 0$, $y < 0$ quadrant. Is the curve concave upward or concave downward? Is this result consistent with what we expect from Fermat's Principle?

Problem 3.2 *Designer Rays*
Use the generalized Snell's Law, (3.1.3), to design a functional dependence of the index of refraction $n(y)$ which allows for
(a) a parabolic ray $y = ax^2$, and
(b) a sinusoidal ray $y = A\sin(kx)$.

Problem 3.3 *Falling Light*
Normally, the index of refraction in the lower earth atmosphere (≤ 10 km) falls off as an exponential $n = 1 + \delta\exp(-\kappa y)$ from mean sea level ($y = 0$) where $\delta = 2.77\ 10^{-4}$ and $\kappa = .105\ \text{km}^{-1} = (9.52\,\text{km})^{-1}$. This relation is a consequence of the Gladstone-Dale Law (Eq. 3.2.2) and the fact that the lower earth atmosphere is roughly isothermal so that $\rho \propto \exp(-mgy/kT)$. A laser beam initially at position $(x = 0, y = h)$ and aligned parallel ($y' = 0$) to the $y = 0$ plane will curve downward in this atmosphere and eventually meet the ground at $(x = L, y = 0)$ as illustrated in figure 3.3.

(a) Show that the first order differential equation governing the ray path is

$$\frac{n(y)}{\sqrt{1 + y'^2}} = n_h$$

where $n_h = n(h)$.

(b) Because the index of refraction $n(y)$ changes only slowly along the ray path, we need only consider leading order corrections to the $y = 0$ level index of refraction n_o, that is,

$$n(y) \approx n_o + n_o' y$$

41

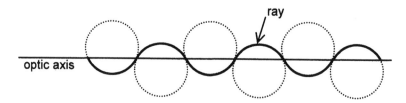

Fig. 3.4. Light ray in a fiber optic

and

$$n(y)^2 \approx n_o^2 + 2n_o n_o' y$$

where the subscript o indicates evaluation at $y = 0$. Therefore $n_o' = (\partial n/\partial y)_o = -\kappa\delta$. Similarly,

$$n_h = 1 + \delta \exp(-\kappa h) \approx 1 + \delta - \delta\kappa h.$$

Make use of these expressions, separate variables in the equation governing the ray path, and integrate the result to arrive at

$$L = \sqrt{\frac{2n_o h}{\kappa\delta}}.$$

Here we have also employed the very good approximations $n_o + n_h \approx 2$, $n_o \approx 1$, and $n_h \approx 1$.

(c) Show that when $h = 1$ meter, $L = 8.3$ km.

Problem 3.4 *Semicircular Rays*
The context is section 3.3. Given that the index of refraction n in a cylindrical fiber optic is the function $n(r) = n_o/(1 + \lambda r)$, show that the meridional rays with $\theta = $ constant form a connected series of circular segments in the $r-z$ plane. See figure 3.4.

Problem 3.5 *Duffing's Equation*
Again the context is section 3.3. Consider a cylindrical gradient index lens with index $n(r) = n_o(1 - \alpha r^2)$.

(a) Find the second order differential equation describing meridional rays passing through the lens in the $\theta = 0$ plane.

(b) Specialize this equation to paraxial rays which remain almost parallel to the lens symmetry axis, i.e., to rays for which $r' \ll 1$.

(c) Show that for weak index gradients ($\alpha r^2 < 1$) the rays become solutions of Duffing's nonlinear wave equation, $r'' + 2\alpha r(1 + \alpha r^2) = 0$ (through first order in αr^2), or harmonic functions with wavelength $\lambda = 2\pi/(2\alpha)^{1/2}$ (through zeroth order in αr^2).

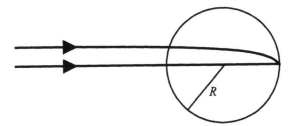

Fig. 3.5. Luneberg sphere

Problem 3.6 *Helix*

Find the form of the cylindrically symmetric index of refraction $n(r)$ for which the light rays are helices with $r = r_o$. (Answer: $n(r)^2 = a + b/r^2$ where a and b are constants.)

Problem 3.7 *Exponential Spiral*

Consider a spherically symmetric material in which the index of refraction decreases with radius r as $n(r) = k/r$.

(a) Show that the light rays are exponential spirals, i.e., $r \propto \exp(a\theta)$ in a $\phi = $ constant plane passing through the center.

(b) Under what condition does the light ray occupy a closed circular path?

Problem 3.8 *Luneberg Lens*[4]

Show that the Luneberg Lens, a sphere of radius R and index $n(r) = (2 - r^2/R^2)^{1/2}$, focuses parallel rays striking the sphere to a point at the back edge of the lens. See figure 3.5.

Problem 3.9 *Time Evolution Equations*

The context is section 3.4. Show that the Euler-Lagrange equations implied by Fermat's Principle when time is used as a parameter can be expressed in vector form as

$$v\nabla \mathbf{n} = \frac{d}{dt}\left[\frac{n\mathbf{v}}{v}\right]$$

where $v = \sqrt{x'^2 + y'^2 + z'^2}$ is the local light speed.

[4]Described by M. Born and E. Wolf in *Principle of Optics* (New York: Pergamon, 1980), p. 146. Apparently, its original source is unpublished notes by R. K. Luneberg titled *Mathematical Theory of Optics* (Brown University, Providence, R. I., 1944), p. 213.

Problem 3.10 *Tautological First Integral*

Show that the first integral corresponding to the fact that the integrand (3.4.1) of the propagation time integral in parametric form does not depend on the independent variable Ω is tautological, that is, reduces to a statement logically equivalent to $x = x$.

Least Potential Energy

> It is reasonable that each kind of body should be
> carried to its own place.
> —Aristotle, *Physics*, 384–322 B.C.E.

MUCH of the world around us—the structures in which we live, a book lying on the table, a spider web on a still day—is at rest and occupies a position of minimum potential energy. Other objects move toward or around a state of least potential energy. The Principle of Least Potential Energy describes the equilibrium position of an object or system of objects. Like other physical principles the Principle of Least Potential Energy correlates many disparate phenomena.

4.1 PRINCIPLE OF LEAST POTENTIAL ENERGY

Imagine putting a small ball inside a dish. If we gently place it at the very bottom, the ball stays put; otherwise, it rolls toward the bottom. We know the result without actually doing the experiment because everyday experience presents us with many closely related phenomena. We know generally, if not precisely, something of the conditions accompanying rest and motion.

Assume the dish is symmetrically shaped about a central axis. The only resting place is at center-bottom where the potential is minimum (fig. 4.1.1a). In analytical language we could describe the dish's height z as a function of distance r from its symmetry axis by $z(r)$. Then the ball's gravitational potential energy is given by

$$U = mgz(r), \qquad (4.1.1)$$

and the ball's possible resting place, at $r = 0$, is determined by the condition

$$\frac{dU(r)}{dr} = 0. \qquad (4.1.2)$$

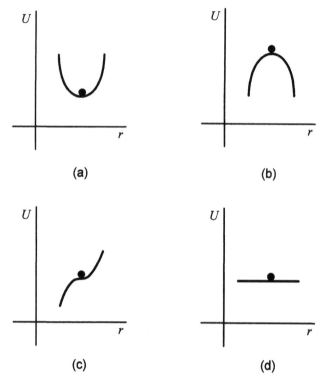

Fig. 4.1. Surfaces with (a) a stable equilibrium, (b) an unstable equilibrium, (c) an inflection point, and (d) a region of neutral equilibrium

Note, also, that

$$\frac{dU^2(r)}{dr^2}\bigg|_{r=0} > 0. \qquad (4.1.3)$$

According to these two equations, $r = 0$ is both a stationary value and a local minimum of $U(r)$.

Every fundamental interaction and many derivative ones have their own potential function. And we must know these functions in order to take advantage of the Principle of Least Potential Energy.[1] Although

---

[1] If, instead, we know the associated conservative force field $\mathbf{F}$ describing these interactions, we can determine the potential energy function from $U(\mathbf{r}) = U(\mathbf{r}_o) - \int_{r_o}^r \mathbf{F} \cdot d\mathbf{r}$ where $U(\mathbf{r}_o)$ is the energy at an arbitrary fiducial point r_o.

Fig. 4.2. Simple system seeking a position of least potential energy

elementary examples treat only of single particles, we will also be concerned with systems composed either of several, discrete, interacting parts or of an extended, continuous, distribution of parts. Such a composite system is in a state of rest when none of the coordinates describing the system change in time. According to the *Principle of Least Potential Energy*, a system in a state of rest in an *inertial reference frame* occupies a stationary value of its potential energy function.

A reference frame rigidly attached to the earth is a good approximation of an inertial reference frame. One of Galileo's (1564–1642) important contributions to the science of mechanics was to argue convincingly that a reference frame moving uniformly with respect to the earth's inertial reference frame is also inertial. Later, with the publication of Newton's *Principia* (1687), it became possible to consistently define an inertial reference frame as one in which Newton's Second Law of Motion, $\mathbf{F} = m\mathbf{a}$, held.

Note that this principle is more properly a *Principle of Stationary Potential Energy*. However, just as we prefer the expression "least time" over "stationary time" to identify Fermat's Principle, we also prefer "least potential" over "stationary potential." Stationary values of a potential energy function are referred to as *equilibria*, that is, places where the net force vanishes. Local minima (fig. 4.1a) are *stable equilibria* because a system displaced to either side of a stable equilibrium moves toward that equilibrium. Local maxima and inflexion points (figs. 4.1b and c) are *unstable equilibria* while finite regions of constant potential (fig. 4.1d) are regions of *neutral equilibrium*. As illustrated in figure 4.2, systems displaced even slightly from an unstable equilibrium move toward a stable one. Local minima of the potential energy function are natural attractors in a world dominated by kinetic friction; for this reason, the Principle of Least Potential Energy is an appropriate name. Problem 4.1 *Loaded Flywheel* describes a system with both stable and unstable equilibria.

This Principle of Least Potential Energy echoes an old idea often incorporated into "common sense" physics. The idea, discussed extensively in Book IV of Aristotle's (384–322 B.C.E.) *Physics*, is that each kind of element (earth, air, fire, or water) has its "natural place." According to the *Physics*, objects leave their natural places only when impelled, say by human agent, and, furthermore undisturbed would return to their natural places. The natural place of every object is determined by its predominate constituent (earth, air, fire, or water). The order achieved when all are in their places is that generally observed: earth below and fire (i.e., the stars) above with water and air intermediately positioned. When "natural places" are identified with potential energy minima, the Principle of Least Potential Energy captures what is most right about this part of Aristotelian physics.

Today we recognize the Principle of Least Potential Energy as a restatement of Newton's first law of motion: "Every body continues in its state of rest, or of uniform motion in a straight line, unless it is compelled to change that state by forces impressed upon it." But the Principle as stated is at the same time less general (comprehending only interactions derivable from a potential energy function) and more general, as we shall see, (applying as well to composite systems naturally incorporating constraints among parts) than Newton's statement of the first law.

Mathematically, the task of finding stationary values is simply one of taking derivatives of a scalar function and solving scalar equations. Higher order derivatives determine stability properties.[2] There is no need to add forces, determine torques, or otherwise do vector algebra. This is one of the characteristic advantages of the variational formulation of physical principles.

---

[2]Stability properties are mathematically determined by the sign of the second derivative of the potential energy function when this does not vanish. When the second derivative does vanish, higher order ones must be consulted. In general, if the first nonvanishing derivative is of even order (e.g., $d^2U/dr^2|_{r_o}, d^4U/dr^4|_{r_o}, d^6U/dr^6|_{r_o}, \ldots$ where r is the independent variable and r_o the stationary value) and its sign is positive, the stationary point is a minimum and the equilibrium stable. Alternatively, if the first nonvanishing derivative is even order and its sign is negative, the stationary point is a maximum and the equilibrium unstable. If the first nonvanishing derivative is of odd order (e.g., $d^3U/dr^3|_{r_o}, d^5U/dr^5|_{r_o}, \ldots$), the stationary point is an inflection point—again, we have instability. Finally, if all the derivatives of U vanish at r_o, the function is constant in a small region surrounding r_o and the region is one of neutral equilibrium.

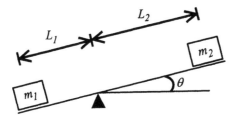

Fig. 4.3. Lever coordinates

4.2 ELEMENTARY EXAMPLES

Spring and Mass. A block of mass m hangs from the end of a vertically suspended massless spring of spring constant k. The spring obeys
Hooke's Law by exerting a force on the block equal to $-kz$ where z is
the displacement of the end of the spring from its unstretched position
at $z = 0$. The total potential energy, gravitational and elastic, is

$$U(z) = mgz + \frac{kz^2}{2}.$$

(4.2.1)

The first derivative of the potential energy, dU/dz, vanishes at

$$z = -\frac{mg}{k},$$

(4.2.2)

while the second derivative

$$\frac{d^2U}{dz^2} = k > 0$$

(4.2.3)

for all values of z. Apparently, there is only one point ($z = -mg/k$) at
which both conditions, $dU/dz = 0$ and $d^2U/dz^2 > 0$, prevail. If this
system were motionless, the mass at the end of the spring would occupy
that point ($z = -mg/k$).

Lever. A lever composed of a pivoted rigid rod supports two masses,
m_1 and m_2, at either end. The masses are fixed on the lever (see fig. 4.3),
but the lever itself is free to rotate around its pivot. Kinetic friction, say,
has brought the lever to rest.

The total gravitational potential energy U of this system is given by

$$U = -m_1gL_1 \sin\theta + m_2gL_2 \sin\theta.$$

(4.2.4)

The necessary condition for a local minimum in $U(\theta)$ is $dU/d\theta = 0$, that is,

$$\frac{dU}{d\theta} = g\cos\theta[-m_1L_1 + m_2L_2] = 0 \qquad (4.2.5)$$

or

$$\theta = \pm\frac{\pi}{2} \qquad (4.2.6)$$

whenever $-m_1L_1 + m_2L_2 \neq 0$. Suppose that $m_2L_2 > m_1L_1$. Then

$$\left.\frac{d^2U}{d\theta^2}\right|_{\theta=+\pi/2} = -g[-m_1L_1 + m_2L_2] < 0 \qquad (4.2.7)$$

and

$$\left.\frac{d^2U}{d\theta^2}\right|_{\theta=-\pi/2} = g[-m_1L_2 + m_2L_2] > 0. \qquad (4.2.8)$$

Therefore, given the condition $m_2L_2 > m_1L_1$, $\theta = -\pi/2$ is a minimum of the potential energy and a position of stable equilibrium, while $\theta = +\pi/2$ is a potential maximum and a position of unstable equilibrium. Thus the Principle of Least Potential Energy claims that the resting lever will be vertical with the heavy mass at the bottom ($\theta = -\pi/2$). Of course, when $m_2L_2 = m_1L_1$, $U(\theta) = 0$ for all θ, all the derivatives of $U(\theta)$ vanish, and $U(\theta)$ is everywhere flat. Then the lever can hang in neutral equilibrium at any value of θ.

4.3 CONSTRAINTS

A system (e.g., spring-mass-earth or lever-mass-earth) is composed of related parts. Sometimes these relations are expressed verbally ("The masses are fixed on the lever") and sometimes algebraically ($L_1 = \text{const.}$, $L_2 = \text{const.}$). In the lever example these relations are consequences of static frictional forces. In another context other forces maintain the rigidity of a beam or the incompressibility of a fluid. Variational methods allow us to account for the effect of these "constraining" forces through kinematical relationships called *constraints* or *auxiliary conditions* without understanding all there is to know about their cause.

Hydrostatic Balance. Two vertical tubes of incompressible fluid of mass density ρ are connected by a closed channel. The tubes, one with cross-sectional area A_1 and one with A_2, are fitted with piston heads upon which weights of mass m_1 and m_2 are placed. See figure 4.4. The

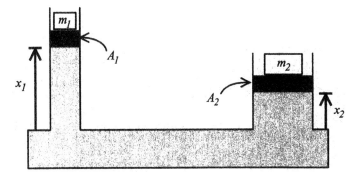

Fig. 4.4. Hydrostatic balance

fluid columns adjust their heights under the influence of the masses and their own weight in order to minimize the system's potential energy.

The gravitational potential energy of this system is, within a constant, given by

$$U = m_1 g x_1 + \frac{1}{2} \rho g A_1 x_1^2 + m_2 g x_2 + \frac{1}{2} \rho g A_2 x_2^2. \qquad (4.3.1)$$

Since the fluid is incompressible, the two variables x_1 and x_2 are not independent but linked by a constant fluid volume V, that is,

$$V = x_1 A_1 + x_2 A_2. \qquad (4.3.2)$$

Using the constraint (4.3.2) to eliminate x_2 from equation (4.3.1) we arrive at

$$\begin{aligned} U &= m_1 g x_1 + \frac{1}{2} \rho g A_1 x_1^2 + m_2 g \left(\frac{V - x_1 A_1}{A_2} \right) \\ &+ \frac{1}{2} \rho g A_2 \left(\frac{V - x_1 A_1}{A_2} \right)^2. \end{aligned} \qquad (4.3.3)$$

The condition for stationary potential energy,

$$\left. \frac{dU}{dx_1} \right|_{x_1 = x_{1eq}} = 0, \qquad (4.3.4)$$

leads to

$$m_1 + \rho A_1 x_{1eq} - m_2 \frac{A_1}{A_2} - \rho A_1 \left(\frac{V - x_{1eq} A_1}{A_2} \right) = 0. \qquad (4.3.5)$$

Solving for x_{1eq} we find that

$$x_{1eq} = \frac{\rho V A_1 + m_2 A_1 - m_1 A_2}{\rho A_1 (A_1 + A_2)}, \tag{4.3.6a}$$

and, given the constraint (4.3.2), we also get

$$x_{2eq} = \frac{\rho V A_2 - m_2 A_1 + m_1 A_2}{\rho A_2 (A_1 + A_2)}. \tag{4.3.6b}$$

Since

$$\frac{d^2 U}{dx_1^2} = \rho g A_1 \left(1 + \frac{A_1}{A_2} \right) > 0 \tag{4.3.7}$$

for all x_1, the equilibrium described by (4.3.6a,b) is stable.

An interesting physical interpretation of the equilibrium conditions (4.3.6a,b) exists. After some manipulation these can be shown to imply

$$\frac{g[m_1 + \rho A_1 x_{1eq}]}{A_1} = \frac{g[m_2 + \rho A_2 x_{2eq}]}{A_2}. \tag{4.3.8}$$

Therefore, the equilibrium is achieved by means of balancing the fluid pressure at the bottom of each column. Evidently, a small mass m_1 on a small cross-sectional area A_1 piston can support a much larger mass m_2 provided the cross-sectional area A_2 on which it rests is chosen to meet the condition (4.3.8). This is the idea behind the transmission of force in automobile brakes and other hydraulic systems. We owe the first clear analysis and interpretation of the hydrostatic problem to Blaise Pascal (1623–62). Equation (4.3.8) is an application of *Pascal's Principle* which states that the pressure within the volume of a static fluid depends only upon depth.

4.4 LAGRANGE MULTIPLIERS

In the hydrostatic balance problem of section 4.3 our mathematical task was to find the stationary values of a function $U(x, y)$ when the variables x and y were themselves not independent but related by a constraint equation of the form

$$C(x, y) = \text{constant}. \tag{4.4.1}$$

Formally, our procedure was to solve (4.4.1) for y in terms of x, that is, find $y = f(x)$, and substitute this result into the function $U(x, y)$ giving

$U[x, f(x)]$. Now, the x which makes U stationary satisfies $dU/dx = 0$, which is

$$\frac{dU}{dx} = \frac{\partial U}{\partial x} + \frac{\partial U}{\partial f}\frac{df}{dx} = 0. \tag{4.4.2}$$

While this procedure is often easy to implement, we wish to introduce another, equivalent, method which will also work when either we cannot or do not wish to solve the constraint equation (4.4.1) directly.

According to the method of *Lagrange multipliers*, we form a new function, the *augmented potential energy*, U^*,

$$U^*(x, y) = U(x, y) + \lambda C(x, y), \tag{4.4.3}$$

from $U(x, y)$ and $C(x, y)$. Here λ is a constant Lagrange multiplier.[3] Then we find stationary values of $U^*(x, y)$, treating both x and y as independent variables, so that

$$\frac{\partial U^*}{\partial x} = \frac{\partial U}{\partial x} + \lambda\frac{\partial C}{\partial x} = 0 \tag{4.4.4a}$$

and

$$\frac{\partial U^*}{\partial y} = \frac{\partial U}{\partial y} + \lambda\frac{\partial C}{\partial y} = 0. \tag{4.4.4b}$$

The three equations (4.4.1 and 4.4.4a,b) determine the three unknowns: x, y, and λ. Our problem is solved. This method is straightforwardly generalized to apply when a function of n coordinates is to be made stationary and the n coordinates are constrained by m ($m < n$) auxiliary conditions so that there are m Lagrange multipliers. Then we must solve $n + m$ equations.

That the method of Lagrange multipliers is equivalent to the method of direct substitution and solution is easily observed. We simply recast the constraint (4.4.1) in terms of its solution $y = f(x)$ so that (4.4.1) takes the form

$$C(x, y) = y - f(x) = 0. \tag{4.4.5}$$

Then the augmented potential $U^*(x, y) = U(x, y) + \lambda[y - f(x)]$ and the two conditions (4.4.4a,b) become

$$\frac{\partial U}{\partial x} - \lambda\frac{df}{dx} = 0 \tag{4.4.6a}$$

---

[3]Here, as in many applications, λ is restricted to be a constant parameter. If λ is allowed to be a function of the variables x and y, that is, $\lambda = \lambda(x, y)$, the constraint equation must be written in the form $C(x, y) = 0$.

and

$$\frac{\partial U}{\partial y} + \lambda = 0. \tag{4.4.6b}$$

When λ is eliminated from equations (4.4.6a,b) the result is equivalent to our previous result, equation (4.4.2).

Let's apply the method of Lagrange multipliers to the example of the hydrostatic balance. Its augmented potential is

$$
\begin{aligned}
U^*(x_1, x_2) &= m_1 g x_1 + \frac{1}{2}\rho g A_1 x_1^2 + m_2 g x_2 + \frac{1}{2}\rho g A_2 x_2^2 \\
&\quad + \lambda(x_1 A_1 + x_2 A_2)
\end{aligned}
\tag{4.4.7}
$$

where the constraint $x_1 A_1 + x_2 A_2 = V$ (a constant) plays the role of $C(x_1, x_2) = $ constant. The parameters x_1 and x_2 are now considered independent variables. The equilibrium conditions are $\partial U^*/\partial x_1 = 0$ and $\partial U^*/\partial x_2 = 0$. These yield

$$m_1 g + \rho g A_1 x_{1eq} + \lambda A_1 = 0 \tag{4.4.8a}$$

and

$$m_2 g + \rho g A_2 x_{2eq} + \lambda A_2 = 0 \tag{4.4.8b}$$

which together with the constraint $x_1 A_1 + x_2 A_2 - V = 0$ result in equations (4.3.6a,b) for x_{1eq} and x_{2eq} as well as

$$\lambda = -\frac{m_1 g + m_2 g + \rho V g}{A_1 + A_2}. \tag{4.4.9}$$

However, note that the equilibrium conditions (4.4.8a,b) alone recover the pressure balance equation (4.3.8),

$$\frac{m_1 g + \rho g A_1 x_{1eq}}{A_1} = \frac{m_2 g + \rho g A_2 x_{2eq}}{A_2}. \tag{4.4.10}$$

In this way one can extract useful information from the equilibrium conditions without explicitly solving the constraint equation.

In this problem the independent variables on which the potential energy depends specifies the system's spatial arrangement. This need not always be the case. More generally, the potential energy U and the constraint function C are functions of abstract coordinates $(q_1, q_2, \ldots q_n)$ whose only purpose is to uniquely specify the potential energy of the system. These coordinates could be, as in problem 4.4 *Electrostatic Energy*, the charges on a set of capacitors.

4.5 CATENARY

An inextensible but flexible chain (*catenary* in Latin) or rope of specified length L hangs between two fixed points, (x_1, y_1) and (x_2, y_2), under the influence of gravity in the x–y plane. What is the curve $y(x)$ describing the chain's shape? In answering this question we apply several of the principles and methods thus far presented: the Principle of Least Potential Energy, the variational calculus, and Lagrange multipliers.

The chain's gravitational potential energy,

$$U = \rho g \int y(x)\, ds = \rho g \int_{x_1}^{x_2} y\sqrt{1 + y'^2}\, dx, \qquad (4.5.1)$$

must be extremized subject to the constraint that its length,

$$L = \int ds = \int_{x_1}^{x_2} \sqrt{1 + y'^2}\, dx, \qquad (4.5.2)$$

be a constant value. In (4.5.1) ρ is the chain's constant mass per unit length while equation (4.5.2) is a constraint on the function $y(x)$. Because (4.5.2) is not an algebraic equation, we do not have the option of solving it for information about the form of $y(x)$. Rather, we must incorporate (4.5.2) into the variational problem via the method of Lagrange multipliers. We form the augmented potential,

$$U^* = \rho g \int_{x_1}^{x_2} y\sqrt{1 + y'^2}\, dx + \lambda \int_{x_2}^{x_1} \sqrt{1 + y'^2}\, dx, \qquad (4.5.3)$$

or, equivalently,

$$U^* = \int_{x_1}^{x_2} (\rho g y + \lambda)\sqrt{1 + y'^2}\, dx \qquad (4.5.4)$$

where λ is the Lagrange multiplier. Since the integrand $f(y, y')$ of the integral U^* is not a function of the integration variable x, a first integral of the form $f - y'(\partial f/\partial y') = D$ (a constant) follows from the Euler-Lagrange equation. After some rearrangement this first integral becomes

$$\frac{(\rho g y + \lambda)}{\sqrt{1 + y'^2}} = D. \qquad (4.5.5)$$

Solving for y', reducing the result to quadratures, and integrating we find that

$$y(x) = -\frac{\lambda}{\rho g} + \frac{D}{\rho g} \cosh\left(\frac{\rho g x}{D} + C_1\right) \qquad (4.5.6)$$

over a domain for which $y'(x) > 0$. Therefore, the catenary is a hyperbolic cosine. The parameters λ, D, and C_1 are chosen so that the length is L and the endpoints (x_1, y_1) and (x_2, y_2). See, for example, problem 4.5 *Symmetric Catenary*.

Constraints which, like equation (4.5.2), take the form of a definite integral are called *isoperimetric constraints* after the first recorded problem of this type, *Dido's problem* (problem 4.6), in which the maximum area enclosed by a given perimeter is sought.[4]

4.6 NATURAL BOUNDARY CONDITIONS

Thus far we have sought extremizing or true functions $y(x)$ on the interval $x_1 \leq x \leq x_2$ from among a comparison set $Y(x) = y(x) + \varepsilon \eta(x)$ given the defining condition that true functions make the integral

$$\int_{x_1}^{x_2} f(y, y', x) \, dx \tag{4.6.1}$$

stationary. In section 2.2 (see equation 2.2.7) we showed that these true functions are those which make

$$\int_{x_1}^{x_2} \eta(x) \left[\left(\frac{\partial f}{\partial y} \right) - \frac{d}{dx} \left(\frac{\partial f}{\partial y'} \right) \right] dx$$
$$+ \frac{\partial f}{\partial y'} \bigg|_{x_2} \eta(x_2) - \frac{\partial f}{\partial y'} \bigg|_{x_1} \eta(x_1) = 0 \tag{4.6.2}$$

for arbitrary variations $\eta(x)$. When $y(x)$ is fixed on the interval endpoints, that is, $y(x_1) = y_1$ and $y(x_2) = y_2$, and, consequently, the variation $\eta(x)$ vanishes there, that is, $\eta(x_1) = \eta(x_2) = 0$, then (4.6.2) reduces to the Euler-Lagrange equation

$$\frac{\partial f}{\partial y} - \frac{d}{dx} \left(\frac{\partial f}{\partial y'} \right) = 0. \tag{4.6.3}$$

Since differential equations of this form are second order, that is, contain derivatives y'' but no higher, we have already employed exactly the right

---

[4]Dido was the legendary queen of Carthage in Virgil's *Aeneid* who obtained the grant of so much land as might be surrounded by an oxhide. Naturally, she encircled the land with a rope made of strips of the hide joined together.

number of boundary conditions [two: $y(x_1) = y_1$ and $y(x_2) = y_2$] necessary to completely determine a solution.

It is, however, a simple matter to conceive of variational problems whose solution is a curve that is fixed at only one end of the interval so that, say, $\eta(x_1) = 0$ but $\eta(x_2)$ is not specified. For example, what is the shortest distance between the point (x_1, y_1) and the line $x_2 = $ constant? Here we look for a curve $y(x)$ which makes stationary an integral of form (4.6.1) with integrand

$$f(y, y', x) = \sqrt{1 + y'^2}. \tag{4.6.4}$$

The value of $y(x)$ at $x = x_2$ is not known beforehand so that $\eta(x_2)$ does not necessarily vanish. Yet, inspecting (4.6.2), we see that if in addition to the imposed boundary condition,

$$\eta(x_1) = 0, \tag{4.6.5a}$$

we have a second boundary condition

$$\left. \frac{\partial f}{\partial y'} \right|_{x_2} = 0, \tag{4.6.5b}$$

then the desired function $y(x)$ is also a solution of the Euler-Lagrange equation (4.6.3).

In this example the Euler-Lagrange equation then reduces to a first integral,

$$\frac{y'}{\sqrt{1 + y'^2}} = C_1, \tag{4.6.6}$$

while condition (4.6.5b) becomes

$$\left. \frac{y'}{\sqrt{1 + y'^2}} \right|_{x_2} = 0. \tag{4.6.7}$$

These results lead to a curve for which $y'(x) = 0$ everywhere; that is, as expected, the curve is a straight line. Apparently, the boundary condition (4.6.5b), while not imposed in the problem's initial description, follows naturally from its variational nature. For this reason equation (4.6.5b) as well as

$$\left. \frac{\partial f}{\partial y'} \right|_{x_1} = 0, \tag{4.6.8}$$

57

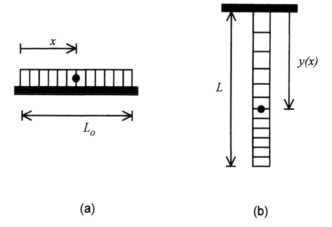

(a) (b)

Fig. 4.5. (*a*) loaded column with no stretching; (*b*) hanging loaded column showing differential stretching

when applicable, are called *natural boundary conditions*. Thus, if otherwise unspecified, choose natural boundary conditions; they along with the Euler-Lagrange equation make the integral (4.6.1) stationary.

Natural boundary conditions are especially useful when the integrand of the integral to be extremized contains higher than first-order derivatives of the unknown function. In each case exactly the right number of natural boundary conditions follows from a mathematical statement of the variational principle. In no kind of application is this better illustrated than in finding the equilibrium configuration of an elastic structure. See problem 4.7 *Loaded Beam*.

4.7 VERTICALLY HANGING ELASTIC COLUMN

A rope, chain, column, "Slinky," or other long, uniform, one-dimensional elastic structure hangs vertically from one end (fig. 4.5). It supports its own weight by stretching differentially so that the elastic restoring force at any one point in the rope balances the weight of the underhanging portion. Obviously, the column stretches more near its top than near its bottom. By hypothesis, the column is resting and in stable equilibrium.

Let the unstretched length of a differential sized segment of the column be dx. The coordinate x designates the displacement of a particular material point from the column's upper end when the column is unstretched (fig. 4.5a). When stretched (or compressed) the length of this segment becomes dy and the position $y(x)$ designates the displacement of the material point from the end of the stretched column (fig. 4.5b). Our task is to find the function $y(x)$ implied by the Principle of Least Potential Energy given that the column contains both elastic and gravitational potential energy.

A differential segment stores elastic potential energy dU_e in the amount

$$dU_e = \frac{\kappa}{2}(dy - dx)^2 \tag{4.7.1}$$

where κ is the "spring constant" of the differential segment. The segment spring constant κ is itself a function of its unstretched length dx; the longer the segment, the smaller the spring constant in inverse proportion. Thus

$$\kappa \, dx = kL_o = E \text{ (a constant)} \tag{4.7.2}$$

where L_o is the unstretched length of the whole column, k is the column's spring constant, and E is the material's *modulus of elasticity*.[5] Using this notation equation (4.7.1) becomes

$$U_e = \int_0^{L_o} \frac{E}{2}(y' - 1)^2 \, dx. \tag{4.7.3}$$

The gravitational potential energy

$$U_g = -g\rho_o \int_0^{L_o} y \, dx \tag{4.7.4}$$

where ρ_o is the constant mass per unit length of the unstretched column. [Note that we could have written the integral (4.7.4) in terms of the variable mass per unit length $\rho(y)$ of the stretched column and integrated over y. But $\rho(y)dy = \rho_o dx$ because segment mass is conserved.] The minus sign in (4.7.4) arises from the fact that positive values of y denote downward displacements and thus smaller values of the potential energy. The total potential energy $U_e + U_g$ of the hanging column is

$$U = \int_0^{L_o} \left\{ \frac{E}{2}(y' - 1)^2 - \rho_o g y \right\} dx. \tag{4.7.5}$$

[5]The modulus of elasticity E is related to Young's constant Y by $Y = AE$ where A is the cross-sectional area of the column. Here A is assumed to remain constant while the column stretches or compresses.

The condition,

$$\int_0^{L_o} \rho_o \, dx = M \text{ (a constant)}, \tag{4.7.6}$$

does not directly restrict the function $y(x)$ since it can be integrated to yield

$$\rho_o = M/L_o. \tag{4.7.7}$$

Thus equation (4.7.7) restricts only the parameters ρ, M, and L_o. We impose the boundary condition $y(0) = 0$ while the value of $y(L_o)$ follows from the natural boundary condition at $x = L_o$,

$$\left. \frac{\partial f}{\partial y'} \right|_{x=L_o} = 0. \tag{4.7.8}$$

Since the integrand

$$f = \frac{E}{2}(y' - 1)^2 - \rho_o g y, \tag{4.7.9}$$

the natural boundary condition, (4.7.8), reduces to

$$y'(L_o) = 1. \tag{4.7.10}$$

An obvious interpretation of (4.7.10) is that the lower end of the column is unstretched. The Euler-Lagrange equation corresponding to making stationary the potential energy, (4.7.5), is

$$Ey'' = -\rho g. \tag{4.7.11}$$

Given the imposed, $y(0) = 0$, and natural, $y'(L_o) = 1$, boundary conditions, the solution of (4.7.11) is the quadratic

$$y(x) = \left(1 + \frac{\rho g L_o}{E}\right) x - \left(\frac{\rho g}{2E}\right) x^2. \tag{4.7.12}$$

The relation between the stretched column length L and unstretched length L_o is now readily found. Since $y = L$ when $x = L_o$, the plausible, and possibly familiar, result

$$L = L_o + \frac{Mg}{2k} \tag{4.7.13}$$

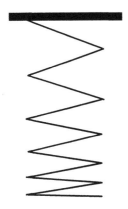

Fig. 4.6. Differential stretching of a massive "Slinky"

follows immediately from (4.7.12) and the relation $\rho = M/L_o$. The total increase in column length, $Mg/2k$, is exactly half the increase in length of a massless spring of constant k with a mass M hanging from its end.

All objects, artificial and natural, are elastic to some degree and so when suspended exhibit the quadratic differential stretching described. This stretching is manifest in a "Slinky" (fig. 4.6) where the coil number plays the role of coordinate x and hidden in a hanging rope and in one of those rock formations suspended from the ceiling of a cave called a stalactite. Even more commonly realized and similarly analyzed is the massive column supported at the bottom.

CHAPTER 4 PROBLEMS

Problem 4.1 *Loaded Flywheel*
Two point masses M and m are fixed on the rim of a massless wheel of radius R. Their angular separation is ϕ_o. See figure 4.7.

(a) Use the circular polar coordinate θ to construct the potential energy function $U(\theta)$. Find the two positions of equilibrium. Check your answer by making sure that $\theta = \phi_o/2$ when $m = M$, and that $\theta \to 0$ as $m \to 0$.

(b) Show that one of the positions is a stable equilibrium while the other is unstable.

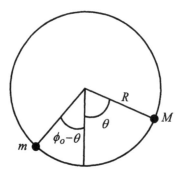

Fig. 4.7. Loaded flywheel

Problem 4.2 *Inclined Plane*

Two masses m_1 and m_2 are connected with a nonelastic string over a frictionless pulley attached to a plane making an angle θ with the horizontal as shown in figure 4.8.

(a) Find the relationship among m_1, m_2, and θ when this system is in equilibrium.

(b) Show that this equilibrium is neutral.

Problem 4.3 *Cantilever Model*

Two identical beams of relaxed length ℓ project from a vertical wall, meet, and support a mass m. They are free to rotate at the wall and stretch or compress but not bend. From the Principle of Least Potential Energy find the distance δ that the upper beam stretches and the distance ε that the lower beam compresses when a mass m is suspended from the cantilever end as shown. Thus the two variables ε and δ completely specify the system configuration. Assume the beam's elastic potential energy is $k\beta^2/2$ when stretched (or compressed) an amount β where the beam "spring constant" k depends upon beam material, relaxed length,

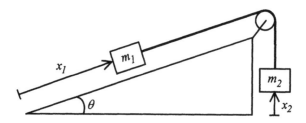

Fig. 4.8. Connected masses on an inclined plane

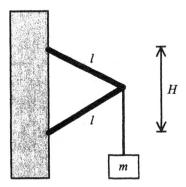

Fig. 4.9. Cantilever model

and cross-sectional area. See figure 4.9. [Hint: The total potential energy is the sum of the elastic and gravitational potential energies. Express the gravitational potential energy of this system as proportional to $(\ell - \varepsilon) \cos \theta$ where θ is the angle the lower support makes with the vertical. Then eliminate $\cos \theta$ with the law of cosines and solve the necessary conditions for equilibrium.]

[Answer: $\varepsilon = \ell m g/(Hk + mg)$ and $\delta = \ell m g/(Hk - mg)$.]

Problem 4.4 *Electrostatic Energy*

The potential energy associated with a capacitor of capacitance C_i holding charge Q_i is $Q_i^2/2C_i$. Find, using the Principle of Least Potential Energy, the charge Q_i on each capacitor of capacitance C_i in terms of the capacitances $C_1, C_2, \ldots C_N$ and the total charge Q on one side of the circuit when the capacitors are connected in parallel. The circuit is shown in figure 4.10. The total potential energy U of the circuit is given by

$$U = \sum_{i=1}^{N} Q_i^2/(2C_i).$$

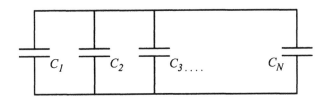

Fig. 4.10. Capacitors in parallel

Use a Lagrange multiplier to incorporate the constraint

$$\sum_{i=1}^{N} Q_i = Q \text{ (a constant)}$$

into the variational process.

Problem 4.5 *Symmetric Catenary*
The context is section 4.5. A symmetrical catenary of length L is hung from the symmetrically located points $(-k, h)$ and (k, h) which, in turn, are adjusted so that the catenary is flat at the origin (that is, $y'(0) = 0$).

(a) Given that the catenary is a function which makes stationary the augmented potential

$$U^* = \int_{-k}^{k} (\rho g y + \lambda)\sqrt{1 + y'^2}\, dx$$

subject to the constraint

$$L = 2\int_{-k}^{k} \sqrt{1 + y'^2}\, dx,$$

show that the form of the symmetric catenary is given by

$$y(x) = \left(\frac{\lambda}{\rho g}\right)[\cosh(\rho g x/\lambda) - 1]$$

where integration constants have been eliminated by applying the symmetric boundary conditions.

(b) Show that the Lagrange multiplier λ is related to the catenary length L and height h by

$$h^2 + \frac{2h\lambda}{\rho g} = \frac{L^2}{4}.$$

Problem 4.6 *Dido's Problem*
A plane figure (4.11a) is bounded by the straight line $y = 0$ and a curve $y(x)$ of length $L = \int_{x_1}^{x_2} \sqrt{1 + y'(x)^2}\, dx$.

(a) For a fixed L show that the curve $y(x)$ which maximizes the area $A = \int_{x_1}^{x_2} y(x)\, dx$ of the figure is an arc of a circle $y(x) = \sqrt{\lambda^2 - a^2} - \sqrt{\lambda^2 - x^2}$ where the Lagrange multiplier λ is the radius and $2a$ is the span of the circular arc (fig. 4.11b).

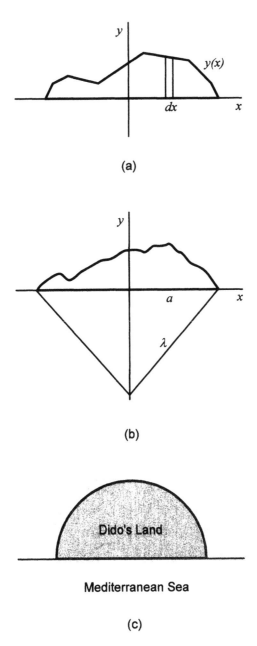

Fig. 4.11a,b,c. Dido's problem

(b) Show that the relationship between the length L, the Lagrange multiplier λ, and the span $2a$ is given by

$$\frac{a}{\lambda} = \sin\left(\frac{L}{2\lambda}\right)$$

so that a is limited to values such that $a \leq \lambda$ where $0 \leq L/2\lambda \leq \pi/2$.

(c) Using the integral defining the bounded area A and the expression derived in part (b) derive the functional dependence

$$A(L, \lambda) = \frac{\lambda}{2}[L - \lambda \sin(2L/\lambda)].$$

(d) Given that $0 \leq L \leq \lambda\pi$ [from part (b)] argue that the area A is largest at the upper boundary of this interval where $\lambda = L/\pi$, that is, when the bounded area is a semicircle (fig. 4.11c).

Problem 4.7 *Loaded Beam*
The context is sections 2.2 and 4.6. The integrand $f(y, y', y'', x)$ of an integral to be extremized contains second order derivatives of the unknown function $y(x)$.

(a) Show that the function $y(x)$ extremizing this integral satisfies the Euler-Lagrange equation

$$\frac{\partial f}{\partial y} - \frac{d}{dx}\left(\frac{\partial f}{\partial y'}\right) + \frac{d^2}{dx^2}\left(\frac{\partial f}{\partial y''}\right) = 0.$$

(b) This is a fourth order differential equation. Find the equation's four natural boundary conditions.

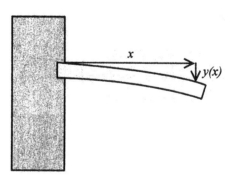

Fig. 4.12. Loaded elastic beam

A massive beam is attached at one end to a wall as shown in figure 4.12. For small total deflections such that $y(L_o) \ll L_o$ the elastic potential energy U_e stored in the drooping beam is given by

$$U_e = \frac{k}{2} \int_0^{L_o} (y'')^2 \, dx$$

where L_o is the beam length and k is an elastic constant.

(c) Find, using the results of parts (a) and (b), the curve $y(x)$ which minimizes the total potential energy (gravitational and elastic) of the loaded beam. Two boundary conditions are imposed, $y(0) = 0$ and $y'(0) = 0$, while the others are natural.

Least Action

> It is what I call the principle of the *least quantity of action*, a principle that is so wise and so worthy of the Supreme Being, and to which nature appears to be so constantly subject that she observes it not only in all her changes but still tends to observe it in her permanence.
> —Pierre Louis Moreau de Maupertuis, 1746

THE PRINCIPLE OF LEAST ACTION determines the geometry of particle trajectories by extremizing the trajectory's action. Because the action of a particle trajectory is mathematically analogous to the optical path length of a light ray, the Principle of Least Action is analogous to Fermat's Principle.

5.1 MAUPERTUIS

A ball rolls on a horizontal surface in a straight line without loss or increase of speed toward a discontinuity. There the ball slows abruptly and continues in a different straight line path. Given that the ball starts at a particular position (x_1, y_1) in the $x-y$ plane, ends at another position (x_2, y_2), and has a trajectory composed of two straight line paths, can we choose the location of the trajectory intercept with the velocity discontinuity (at $x = 0$, $y = ?$) on the basis of minimizing or maximizing a certain quantity? What is that quantity? Figures 5.1a and b remind us that the ball-velocity discontinuity interaction is the mechanical analog of the optical phenomena of refraction. Yet Fermat's Principle does not govern the ball. We find, say with a marble and piece of cardboard folded along two parallel lines, that the marble's trajectory inclines away from the normal when slowing down rather than toward the normal as required by Fermat's Principle. Apparently, the marble takes neither the shortest path (an unbroken straight line) nor the path of least time.

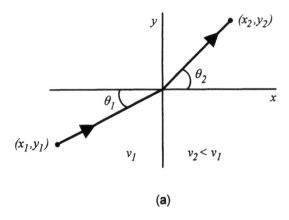

(a)

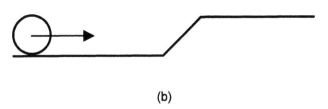

(b)

Fig. 5.1. (*a*) Trajectory of a rolling ball which slows down when crossing the *y*-axis; (*b*) mechanical model of velocity discontinuity

The French mathematician Maupertuis identified the *action* as the quantity minimized by the actual trajectory of a mechanical system. The action of a single particle is the product of its mass m, speed v, and displacement d. According to Maupertuis's *Principle of Least Action*, mvd summed over each straight segment of a particle's trajectory is minimized. For the rolling ball, the action

$$A = mv_1\sqrt{x_1^2 + (y - y_1)^2} + mv_2\sqrt{x_2^2 + (y - y_2)^2}. \qquad (5.1.1)$$

In compliance with Maupertuis's principle, A is stationary with respect to variations in the parameter y locating the trajectory-interface intercept. This requirement, that is, $dA/dy = 0$, leads to

$$v_1 \sin\theta_1 = v_2 \sin\theta_2 \qquad (5.1.2)$$

where $\sin\theta_1 = (y - y_1)/\sqrt{x_1^2 + (y - y_1)^2}$. Equation (5.1.2) determines the trajectory y intercept and correctly predicts that the ball's trajectory bends away from the discontinuity normal (i.e., $\theta_2 > \theta_1$ in fig. 5.1) when the ball slows down. In the language of Newtonian mechanics a force normal to the discontinuity diminishes the x component of the velocity while the y component of the particle momentum is conserved. Equation (5.1.2) also actually minimizes and not only makes stationary the action (5.1.1) (see problem 5.1 *Minimum Action*). But the Principle of Least Action, like Fermat's Principle, requires only that the true trajectory occupy a stationary value of the action rather than a minimum or maximum value. Therefore, the Principle of Least Action is, properly speaking, a *Principle of Stationary Action*.

Maupertuis claimed that his principle applied not only to rolling balls but also to light rays and that, consequently, equation (5.1.2) was, indeed, Snell's Law. The attempt to incorporate ray optics into mechanics was plausible in Maupertuis's time. Newton, sixty years earlier, and Descartes, before Newton, had similar ambitions.[1] Ultimately these ambitions were frustrated. Not until 1850 did Foucault obtain convincing experimental evidence that the refraction of light was governed instead by the Principle of Least Time, that is, by equation (1.2.4),

$$\frac{\sin\theta_1}{v_1} = \frac{\sin\theta_2}{v_2}. \tag{5.1.3}$$

In the meantime Maupertuis promoted the Principle of Least Action.[2] He believed it reflected the simplicity and economy of the Creator-God's work and boldly claimed to have deduced the principle from even more general postulates. Unfortunately, Maupertuis never gave the principle

[1] For a discussion of the several, contradictory models of light, advanced by Descartes, see either the historical survey *The Nature of Light* by Vasco Ronchi (Cambridge, Mass.: Harvard University Press, 1970), pp. 112 ff. or *Theories of Light* by A. I. Sabra (London: Cambridge University Press, 1981), chapter 5. Descartes took as evidence of the particle nature of light the fact that cannonballs shot into water "reflect" and "refract" in a manner analogous to the reflection and refraction of light. "This has been sometimes demonstrated with unfortunate consequences when someone firing guns for fun into the bed of a river has wounded those who were on the other side of the bank." (From the second discourse of Descartes's *Dioptrique*, quoted in Ronchi, p. 117.)

[2] Maupertuis's priority of discovery of the Principle of Least Action has been disputed, but the dispute is generally resolved in his favor. For a brief account, see Yourgrau and Mandelstam, *Variational Principles in Dynamics and Quantum Theory* (New York: Dover, 1979), pp. 22–23.

a precise definition capable of application in any but the most elementary mechanical problems, nor was he an able enough mathematician to adopt the methods of the calculus of variations, then in their infancy, to its service. Yet his advocacy beginning in 1744 initiated a century of fruitful investigation pushed forward by Euler (1707–93), Lagrange (1736–1813), Jacobi (1804–51), and Hamilton (1805–65).

5.2 JACOBI'S PRINCIPLE OF LEAST ACTION

In section 5.1 the speed v of the particle was assumed known as a function of the spatial coordinates. This function was then used to formulate the action of each trajectory in a comparison set of trajectories. In particular, the action A was parameterized by a single variable y and the true trajectory was identified with a solution of $dA/dy = 0$. According to the Principle of Least Action, the true trajectory is the one which makes the action stationary.

The requirement that we know the particle speed v as a function of the spatial coordinates x, y, and z, is generally supplied by a statement of energy conservation

$$T + U(x, y, z) = E \text{ (a constant)} \tag{5.2.1}$$

where T is the particle kinetic energy

$$T = \frac{mv^2}{2} \tag{5.2.2}$$

and $U(x, y, z)$ is assumed given. Combining (5.2.1) and (5.2.2) we find the required functional dependence

$$v(x, y, z) = \sqrt{\frac{2\{E - U(x, y, z)\}}{m}}. \tag{5.2.3}$$

As the particle moves through a differential distance ds its action is incremented by the product $m\, v\, ds$. The total action of a particle trajectory is then the integral

$$A = \int mv(x, y, z)\, ds, \tag{5.2.4}$$

that is,

$$A = \int \sqrt{2m\{E - U(x, y, z)\}}\, ds. \tag{5.2.5}$$

A true trajectory makes this action stationary. While this statement is essentially Maupertuis's Principle of Least Action made precise, it has become known as *Jacobi's Principle of Least Action* or simply *Jacobi's Principle* after C.G.J. Jacobi (1804–51), the mathematician who a century after Maupertuis enjoined this precision.

Typically, the integration variable of the action integral (5.2.5) is one of the coordinates, say x, of the space in which the particle moves. For instance,

$$A = \int_{x_1}^{x_2} \sqrt{2m\{E - U(x, y, z)\}}\sqrt{1 + y'(x)^2 + z'(x)^2}\, dx. \quad (5.2.6)$$

The unknown functions, $y(x)$ and $z(x)$, parametrically describe a curve in coordinate space. Therefore, the action (5.2.6) is a functional, rather than a function, and a true trajectory must be determined with the techniques of the calculus of variations. Note that Jacobi's Principle directly determines the geometry of a particle orbit rather than its dynamics and can be applied only when the particle energy is conserved.

Trajectories of a free particle, called *geodesics*, are easily found by Jacobi's Principle. A particle is free when it moves within a region where the potential U is a constant. Because the speed v of a free particle is constant,

$$A = m v \int ds. \quad (5.2.7)$$

Hence, functions which make the action stationary, in this case, also make the path length between trajectory endpoints stationary. In unconstrained Euclidean three-space geodesics are, of course, straight lines. But, when the path is constrained to lie on some surface, geodesics assume a more complicated form. Geodesics are great circles on the surface of a sphere and are helices on a right cylinder. See problem 5.3 *Geodesic*.

5.3 PROJECTILE TRAJECTORY

Applications of Jacobi's Principle and other variational principles require two steps. The first is to form the definite integral to be made stationary and to use the relevant principle to deduce the Euler-Lagrange differential equation or, equivalently, its first integral. The second step is to solve the differential equation given boundary or initial values. Our primary interest is in the important first step: deducing the differential equation.

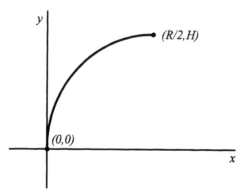

Fig. 5.2. Projectile trajectory with height H
and range R

The first step alone distinguishes the variational approach to mechanics
from the Newtonian.

Let's use Jacobi's Principle to find the shape of a projectile trajectory
near the surface of the earth. According to (5.2.4), a trajectory's action
depends both on the particle speed and path length. From this fact we
argue that a projectile trajectory is concave downward. By curving upward and consequently diminishing its speed, the projectile diminishes
its action; by lengthening its path, the particle increases its action. The
calculus of variations finds the compromise curve with the least possible
action.

Specifically, we seek a trajectory connecting the points (0,0) and
$(R/2, H)$ where the latter is defined to be the point were the trajectory is flat, as illustrated in figure 5.2. Since the near-earth gravitational
potential energy

$$U(y) = mgy \qquad (5.3.1)$$

where y is the height above the earth's surface, the projectile speed
function $v(y)$ is, according to energy conservation (5.2.3), given by

$$v(y) = \sqrt{\frac{2(E - mgy)}{m}}. \qquad (5.3.2)$$

Thus, the corresponding action of a trajectory $y(x)$ is proportional to

$$A = \int_0^{R/2} \sqrt{[E - mgy]}\sqrt{1 + y'^2}\, dx \qquad (5.3.3)$$

where we have chosen x as the independent integration variable. Therefore, the true trajectory is that which satisfies the Euler-Lagrange equation

$$\frac{\partial f}{\partial y} - \frac{d}{dx}\left(\frac{\partial f}{\partial y'}\right) = 0 \tag{5.3.4}$$

where the integrand

$$f(y, y') = \sqrt{[E - mgy]}\sqrt{1 + y'^2}. \tag{5.3.5}$$

Since this integrand is not an explicit function of the independent variable x, the Euler-Lagrange equation (5.3.4) has a first integral

$$f - y'\frac{\partial f}{\partial y'} = D \text{ (a constant)}. \tag{5.3.6}$$

This is equivalent to the first order differential equation

$$\sqrt{\frac{E - mgy}{1 + y'^2}} = D \tag{5.3.7}$$

or to

$$\sqrt{\frac{E - mgy}{1 + y'^2}} = \sqrt{E - mgH} \tag{5.3.8}$$

where we have evaluated the constant D by requiring that $y' = 0$ when $y = H$. In arriving at equation (5.3.8) we have taken the first step.

The second step is to solve the differential equation (5.3.8). We do this by squaring each side of (5.3.8), isolating y', separating the dependent variable y from the independent variable x, and integrating. The result of this integration,

$$\int_0^y \frac{dy}{\sqrt{H - y}} = \sqrt{\frac{mg}{E - mgH}} \int_0^x dx, \tag{5.3.9}$$

is the parabola

$$\frac{y(x)}{H} = \frac{4x}{R}\left(1 - \frac{x}{R}\right) \tag{5.3.10}$$

where we have required that $y = H$ when $x = R/2$. For a similar example, see problem 5.4 *Orbit Shapes*.

5.4 OPTICS AND MECHANICS

Because Fermat's Principle is formally equivalent to Jacobi's Principle, much of the mathematics developed for ray optics also applies to single particle mechanics. Let's expose the terms of this mathematical equivalence. Recall, though, that the two principles are physically distinct; massive particles do not obey Fermat's Principle nor do light rays obey the Principle of Least Action.[3]

Fermat's Principle claims that a true ray makes the optical path length stationary, while Jacobi's Principle claims that a true particle trajectory makes the action stationary. The optical path length cT and the action A are given, respectively, by

$$cT = \int \frac{ds}{v} \qquad \text{and} \qquad A = \int m v \, ds \qquad \text{(5.4.1a, b)}$$

where in the first equation, (5.4.1a), v is the light velocity and in the second, (5.4.1b), v is the particle velocity. Equations (5.4.1a,b) are, however, mere mnemonics—easy to remember but lacking in definition. In order to define these integrations both an integration variable and the functional dependence of the velocities must be specified. These dependences in each case are

$$v(x, y, z) = \frac{c}{n(x, y, z)} \qquad \text{(5.4.2a)}$$

and

$$v(x, y, z) = \sqrt{\frac{2\{E - U(x, y, z)\}}{m}}, \qquad \text{(5.4.2b)}$$

where the functions $n(x, y, z)$ and $U(xy, z)$ are givens which specify the physics of the situation. Substituting equations (5.4.2a,b), respectively, into (5.4.1a,b), we arrive at

$$cT = \int n(x, y, z) \, ds \qquad \text{(5.4.3a)}$$

and

$$A = \int \sqrt{2m\{E - U(x, y, z)\}} \, ds. \qquad \text{(5.4.3b)}$$

---

[3]Fermat's Principle is not a special case of the Principle of Least Action nor, equivalently, is light composed of tiny projectiles ("corpuscles") as Descartes, Newton, and Maupertuis theorized.

Therefore, when

$$n(x, y, z)^2 \quad \text{is proportional to} \quad E - U(x, y, z), \qquad (5.4.4)$$

the resulting Euler-Lagrange equations, one for light rays and one for particle trajectories, are mathematically identical. While it may be difficult to physically imagine the proportionality (5.4.4), there is, at least formally, a particle trajectory for each ray path and vice versa. For example, the index $n(y) = a + by$ corresponds to a potential $U(y) = c + dy + ey^2$ where a, b, c, d, and e are constants. In this case both the light ray and particle trajectory are cycloids.[4]

CHAPTER 5 PROBLEMS

Problem 5.1 *Minimum Action*
Show that the relation $v_1 \sin \theta_1 = v_2 \sin \theta_2$, that is, equation (5.1.2), actually minimizes the action $A = mv_1\sqrt{x_1^2 + (y - y_1)^2} + mv_2 \sqrt{x_2^2 + (y - y_2)^2}$ by showing that $d^2 A/dy^2 \geq 0$ for all y.

Problem 5.2 *Continuum Limit*
The context is section 5.1. Suppose the speed v of a particle of mass m particle is a function $v(y) = \sqrt{v_o^2 - 2gy}$. Model this function as a series of layered velocity discontinuities through which the particle passes. At each change in the height Δy, the particle speed jumps Δv. See figure 1.5 in chapter 1 for the geometry. Apply the result (5.1.2) to a velocity discontinuity and take the limit $\Delta y/\Delta x \to dy/dx$. Show that this limit is

$$v(y)/\sqrt{1 + (dy/dx)^2} = \text{constant},$$

and that the latter is solved by the parabola $y(x) = -gx^2/(2v_o^2)$.

Problem 5.3 *Geodesic*
Show that, according to the Principle of Least Action, an otherwise free particle constrained to move on the surface of a cylinder of radius $r = $ constant maps out a helix ($\theta = az + b$).

[4] Also, when $n(y) = \sqrt{a + by}$ and $U(y) = ay$, the light path and particle trajectory are parabolas; when $n(y) = a/(1 + by)$ and $U(y) = a + b/(1 + dy)^2$, the light path and particle trajectory are semicircles; and when $n(r) = a/r$ and $U(r) = a + b/r^2$, the light path and particle trajectory are exponential spirals.

Problem 5.4 *Orbit Shapes*

A particle of mass is constrained to move in a plane under the influence of a central potential $U(r)$ where r is the distance of the particle from a fixed force center $r = 0$.

(a) Write the action integral (5.2.5) in (r, θ) plane polar coordinates.

(b) Use the Principle of Least Action to derive the equation

$$\theta = c_1 \int \frac{dr}{\sqrt{r^4[E - U(r)] - c_1^2 r^2}}.$$

When $U(r)$ is known, this integral determines the orbit shape function $r(\theta)$.

(c) Suppose the potential function is a $1/r^2$ attractive gravitational force, that is, $U(r) = -GMm/r$. Integrate the above and show that the orbit shape is a conic section of the form

$$r = \frac{a}{1 + \varepsilon \cos \theta}$$

where a and ε are positive constants. When $\varepsilon < 1$ this equation describes an ellipse, when $\varepsilon = 1$ a parabola, and when $\varepsilon > 1$ a hyperbola.

(d) Find the orbit shape when $U(r) = kr^2/2$.

Problem 5.5 *Spherical Pendulum*

A particle of mass m is constrained to move at the end of a rigid pendulum

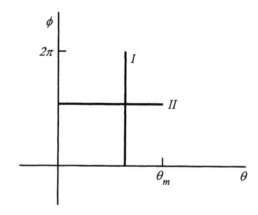

Fig. 5.3. Special orbits of the spherical pendulum: swinging in a vertical plane (I) and moving in a horizontal circle (II)

of length $r = \ell$ under the influence of gravity near the surface of the earth.

(a) Write the action integral (5.2.5) in spherical polar coordinates (θ, ϕ) with $r = \ell$.

(b) Use the Principle of Least Action to find a differential equation whose solution is the function $\phi(\theta)$.

(c) Show that the special orbits $[\theta = \text{const.}, 0 \leq \phi \leq 2\pi]$ and $[0 \leq \theta \leq \theta_m, \phi = \text{const.}]$ are solutions of your differential equation. The limit θ_m is determined by the requirement that the pendulum bob speed $v \propto \sqrt{E - U(\theta)} \geq 0$. See figure 5.3.

Hamilton's Principle—Restricted

> The reader will find no figures in this work. The
> methods which I set forth do not require either
> constructions or geometrical or mechanical
> reasonings: but only algebraic operations, subject
> to a regular and uniform rule of procedure.
> —Lagrange, *Analytic Mechanics*, 1788

HAMILTON'S PRINCIPLE both generalizes and reformulates the Principle of Least Action. The generalization we reserve until chapter 7; here we consider the alternative formulation. The time evolution of a particle trajectory and not only its spatial geometry follows immediately from Hamilton's Principle as does the Principle of Least Potential Energy.

6.1 HAMILTON'S PRINCIPLE

We return to a familiar problem with a familiar geometry: a particle of mass m, initially at a point (x_1, y_1), crosses a planar discontinuity $(x = 0)$ at which the particle's potential energy changes from U_1 to U_2 and then travels to (x_2, y_2). If the particle's total mechanical energy E is conserved, then the particle speed on each side of the discontinuity, v_1 and v_2, is given by

$$v_1 = \sqrt{(2/m)(E - U_1)} \qquad (6.1.1a)$$

for $x < 0$ and

$$v_2 = \sqrt{(2/m)(E - U_2)} \qquad (6.1.1b)$$

for $x > 0$. According to the Principle of Least Action, the action

$$A(y) = mv_1\sqrt{(x_1^2 + (y - y_1)^2)} + mv_2\sqrt{(x_2^2 + (y_2 - y)^2)} \quad (6.1.2)$$

is made stationary for a true trajectory. This requirement leads to

$$v_1 \sin\theta_1 = v_2 \sin\theta_2 \qquad (6.1.3)$$

where the angles θ_1 and θ_2 are defined as in figure 5.1. Such is Maupertuis's analysis as outlined in section 5.1.

However, there is another variational principle from which the same trajectory follows: *Hamilton's Principle*. This principle also uncovers the particle's time evolution and, in its most general form, applies even when the Principle of Least Potential Energy does not. Refer again to figure 5.1. According to the approach of Hamilton's Principle, the particle occupies a position in space-time designated by the coordinates x, y, and t. Thus the particle's initial position in space-time is (x_1, y_1, t_1) and its final position (x_2, y_2, t_2). The intermediate time t ($t_1 < t < t_2$) and y intercept at which the particle crosses the discontinuity are then determined by making stationary, not the action, but rather another quantity

$$S = \int_{t_1}^{t_2} (T - U)\, dt \tag{6.1.4}$$

called *Hamilton's first principal function*.[1] Here the time coordinate t has been favored as the integration parameter. T is the particle kinetic energy $mv^2/2$, U is the particle potential energy, and the spatial coordinates, $x(t)$ and $y(t)$, are functions of time t. As such, the latter parametrically define a trajectory in space-time. Generally, S is a functional of the particle trajectory. Characteristically, Hamilton's Principle, in contrast to the Principle of Least Action, does not insist *a priori* that the comparison class of particle trajectories, $x(t)$ and $y(t)$, contain only those consistent with energy conservation. Rather, the trajectories which conserve energy are determined *a posteriori* by the principle itself.

In the problem of the ball rolling across the discontinuity the integral (6.1.4) reduces to

$$S(y, t) = \frac{mv_1^2}{2}(t - t_1) - U_1(t - t_1) + \frac{mv_2^2}{2}(t_2 - t) - U_1(t_2 - t). \tag{6.1.5}$$

Here the dependence of S on y is implicit in the relation $v_1^2 =$

---

[1] According to Lanczos [*The Variational Principles of Mechanics* (New York: Dover, 1949), p. xxiii], there is no standard name for the quantity S defined in equation 6.1.4. Richard Feynman [*Lectures on Physics* (Reading, Mass.: Addison-Wesley, 1964), vol. II, p. 19-8] incorrectly calls it the "action." The name "Hamilton's first principle function" has the virture of definiteness and is fairly common, if ungainly.

$[x_1^2 + (y - y_1)^2]/(t - t_1)^2$ and $v_2^2 = [x_2^2 + (y - y_2)^2]/(t_2 - t)^2$. Made explicit,

$$S(y, t) = \frac{m[x_1^2 + (y - y_1)^2]}{2(t - t_1)} - U_1(t - t_1)$$

$$+ \frac{m[x_2^2 + (y_2 - y)^2]}{2(t_2 - t)} - U_1(t_2 - t). \qquad (6.1.6)$$

The unknown coordinates y and t which make S stationary satisfy

$$\frac{\partial S}{\partial y} = 0 \qquad (6.1.7a)$$

and

$$\frac{\partial S}{\partial t} = 0. \qquad (6.1.7b)$$

Given (6.1.6) these are equivalent to

$$\frac{y - y_1}{t - t_1} = \frac{y_2 - y}{t_2 - t} \qquad (6.1.8a)$$

and

$$\frac{m v_1^2}{2} + U_1 = \frac{m v_2^2}{2} + U_2, \qquad (6.1.8b)$$

respectively. (See problem 6.1 *Redux*.) The former, (6.1.8a), reproduces (5.1.2) or (6.1.3), that is, $v_1 \sin \theta_1 = v_2 \sin \theta_2$, the "Snell's Law" of particle dynamics. The latter, (6.1.8b), is, of course, energy conservation.

To summarize: Hamilton's Principle states that a true particle trajectory renders Hamilton's first principal functional S stationary. By relaxing the constraint of energy conservation we can vary S over a larger class of comparison functions. This class is determined by two parameters, y and t, instead of one, y, and therefore two conditions (6.1.8a,b) instead of one (6.1.3) follow from the variation. That one of the conditions recovers energy conservation seems almost miraculous. In fact, this miracle, as we shall see, follows from the carefully chosen form of S.

When the energy of the system is conserved, the physical content of Hamilton's Principle, in this case we called it the *Restricted Hamilton's Principle*, is no more or less than the Principle of Least Action. In the next section we derive the former from the latter. A more general version of Hamilton's Principle originated with the eminent Irish mathematician, Sir William Rowan Hamilton (1805–65) and is the subject of chapter 7.

6.2 DERIVING THE RESTRICTED HAMILTON'S PRINCIPLE

Return to the "mnemonic" form of the single particle action integral,

$$A = \int m \, v \, ds. \tag{6.2.1}$$

Recall that in order to vary (6.2.1) one must, first, specify an integration variable and, second, express the speed v as a function of the spatial coordinates by means of energy conservation. Let's replace these two steps with the mathematically equivalent ones of choosing time as the integration parameter, temporarily relaxing the requirement of energy conservation, and reincorporating energy conservation with a Lagrange multiplier. In doing so we will derive the Restricted Hamilton's Principle from the Principle of Least Action.

With time t as integration parameter (6.2.1) becomes the definite integral

$$A = \int_{t_1}^{t_2} m \, v \frac{ds}{dt} \, dt, \tag{6.2.2}$$

or equivalently,

$$A = \int_{t_1}^{t_2} 2T \, dt \tag{6.2.3}$$

since $ds/dt = v$ and

$$T = \frac{mv^2}{2}. \tag{6.2.4}$$

Now, suppose the comparison class of trajectories, inspected in the process of making the action (6.2.3) stationary, is allowed to include trajectories along which particle energy is not conserved. But we want the trajectory actually chosen in the variational process, the true trajectory, to conserve energy, that is, to have

$$T + U = E \text{ (a constant)}, \tag{6.2.5}$$

at every point along the trajectory. In order to incorporate this expectation as an isoperimetric-style constraint,

$$\int_{t_1}^{t_2} (T + U) \, dt = E(t_2 - t_1) \text{ (a constant)}, \tag{6.2.6}$$

on the action (6.2.3), we employ the method of Lagrange multipliers. The appropriate augmented action

$$A^* = \int_{t_1}^{t_2} \{2T + \lambda(T + U)\} \, dt \tag{6.2.7}$$

where λ is a Lagrange multiplier. We find it convenient to denote the integrand of (6.2.7) by f so that

$$f = T(2 + \lambda) + \lambda U. \tag{6.2.8}$$

In general, three coordinates, say x, y, and z, are necessary to completely describe the position of the particle. Therefore, $U = U(x, y, z)$ and $T = m(\dot{x}^2 + \dot{y}^2 + \dot{z}^2)/2$. The Euler-Lagrange equations following from the integrand (6.2.8) are then

$$\frac{\partial f}{\partial x} - \frac{d}{dt}\left(\frac{\partial f}{\partial \dot{x}}\right) = 0, \tag{6.2.9a}$$

$$\frac{\partial f}{\partial y} - \frac{d}{dt}\left(\frac{\partial f}{\partial \dot{y}}\right) = 0, \tag{6.2.9b}$$

and

$$\frac{\partial f}{\partial z} - \frac{d}{dt}\left(\frac{\partial f}{\partial \dot{z}}\right) = 0 \tag{6.2.9c}$$

where a "dot," as is customary, denotes differentiation with respect to time, e.g. $\dot{x} = \frac{dx}{dt}$. The three functions $x(t)$, $y(t)$, $z(t)$ and the Lagrange multiplier λ are determined by the three Euler-Lagrange equations (6.2.9a,b,c) and the constraint (6.2.5) or (6.2.6). One could, in fact, pause here and prove that (6.2.9a,b,c) and (6.2.5) result in the three components of $m\ddot{\mathbf{x}} = -\nabla U$ and $\lambda = -1$ as requested in problem 6.2 *Equations of Motion.* However, our immediate aim is to solve the set (6.2.9a,b,c) and (6.2.5) for the Lagrange multiplier λ alone. To do so, we do not need all three Euler-Lagrange equations. Rather, we need only the constraint equation (6.2.5) and the first integral following from the lack of dependence of the integrand f on the integration variable t, that is,

$$\dot{x}\frac{\partial f}{\partial \dot{x}} + \dot{y}\frac{\partial f}{\partial \dot{y}} + \dot{z}\frac{\partial f}{\partial \dot{z}} - f = D \text{ (a constant)}. \tag{6.2.10}$$

Given our notation, this first integral may be written as

$$T(2 + \lambda) - \lambda U = D \tag{6.2.11}$$

which is consistent with the constraint (6.2.5), $T + U = E$, if and only if $\lambda = -1$ and $D = E$. The next step is telling. We substitute the required value $\lambda = -1$ into the augmented action A^* of (6.2.7) so that it becomes

$$A^* = \int_{t_1}^{t_2} (T - U)\, dt. \tag{6.2.12}$$

85

In this way we build energy conservation (6.2.5) into the very form of the augmented action. Not only do the equations of motion for $x(t)$, $y(t)$, and $z(t)$ follow from making this A^* stationary but so also does energy conservation follow from a lack of dependence of the integrand, $T - U$, on the integration variable t. We have derived the Restricted Hamilton's Principle from the Principle of Least Action. Interestingly, this derivation has no optical analog. See problem 6.3 *An Optical Lagrangian?*

Apparently, a true trajectory minimizes, maximizes, or otherwise makes stationary the time integral of the difference between the kinetic T and potential U energies. We have called this time integral the augmented action A^* but the latter is identical to Hamilton's first principle function S, that is, to

$$S = \int_{t_1}^{t_2} (T - U)\, dt. \qquad (6.2.13)$$

The integrand, $T - U$, is called the *Lagrangian* and denoted by an L so that

$$L = T - U. \qquad (6.2.14)$$

The corresponding Euler-Lagrange equations, in this context, are called *Lagrange's equations of motion*. These designations honor Joseph Louis Lagrange who derived, by another method, Lagrange's equations of motion before Hamilton's Principle, general or restricted, was known.[2]

Our derivation of the Restricted Hamilton's Principle made use of Cartesian coordinates, but it can easily be generalized to accommodate other particle coordinates (q_1, q_2, and q_3) as well. For a proof, see problem 6.4 *Other Coordinates*. In general, Lagrange's equations of motion are:

$$\frac{\partial L}{\partial q_i} - \frac{d}{dt}\left(\frac{\partial L}{\partial \dot{q}_i}\right) = 0 \qquad (6.2.15)$$

---

[2] It is common to derive Lagrange's equations of motion from Hamilton's Principle and speak of the former as following from the latter. This procession is a logical rather than historical one. Lagrange's life (1736–1813) and Hamilton's (1805–65) were almost disjoint with Lagrange *preceding* Hamilton. In fact, Lagrange did not work from Hamilton's Principle but rather from another form of the Principle of Least Action. According to this formulation, a true particle trajectory makes stationary the action $A = \int_{t_1}^{t_2} 2T\, dt$ where the upper limit t_2 is allowed to vary in order that the energy of each trajectory in the set of comparison trajectories remains the same. Lagrange's equations of motion follow from this formulation but only after a further parameterization of the action integral. See Lanczos, pp. 137–38. Although we will not pursue this elaborate method further, its existence explains how Lagrange arrived at the "Lagrangian" and "Lagrange's equations of motion" before Hamilton's Principle was known.

where $i = 1, 2$, and 3. Newton's Second Law of Motion,

$$m\frac{d\mathbf{v}}{dt} = -\nabla \mathbf{U}, \tag{6.2.16}$$

when the force is conservative follows easily from Lagrange's equations of motion (6.2.15). Simply express the Lagrangian L in terms of Cartesian coordinates,

$$L = \frac{m(v_x^2 + v_y^2 + v_z^2)}{2} - U(x, y, z), \tag{6.2.17}$$

so that $q_1 = x, q_2 = y$, and $q_3 = z$. Although equivalent, the variational approach has certain advantages over the Newtonian for determining the ordinary differential equations which govern particle dynamics. For instance, the Lagrangian L and thus Hamilton's function S may be expressed in terms of any set of coordinates $(q_i, i = 1 \ldots)$ which completely specify the state of a particle. In particular, coordinates may be chosen so as to render either the constraints or the potential energy simply expressed. Once L is written, all else is straightforward mathematical analysis of the simplest kind: taking derivatives of the scalar quantity L. In contrast, Newton's Second Law is easy to write in Cartesian coordinates, but in others the acceleration depends in nonobvious ways on the time derivatives of the coordinate unit vectors.

Finally, we note that when the particle is stationary, $T = 0$ and Hamilton's first principal function S reduces to a function

$$S(x, y, z) = -U(x, y, z)(t_2 - t_1) \tag{6.2.18}$$

of the particle coordinates. Thus the values of x, y, and z which make $S(x, y, z)$ stationary also make $U(x, y, z)$ stationary and the Restricted Hamilton's Principle reduces to the Principle of Least Potential Energy.

6.3 SPHERICAL PENDULUM

Let's exploit the flexibility inherent in Hamilton's Principle and Lagrange's equations of motion. Suppose, for instance, one wants to produce equations of motion for a spherical pendulum, that is, a gravitating mass constrained to move on the surface of a sphere of radius ℓ. The coordinates most suited to the problem are, it seems, spherical polar coordinates r, θ, and ϕ. The radial coordinate r is constrained by the supposed rigidity of the pendulum, that is, $r = \ell$ (a constant).

Because only two coordinates $\theta(t)$ and $\phi(t)$ are necessary to completely determine the pendulum's position, the pendulum is said to have two *degrees of freedom*. The Lagrangian is

$$L = \frac{m(\ell^2\dot{\theta}^2 + \ell^2\sin^2\theta\dot{\phi}^2)}{2} - mg\ell(1 - \cos\theta), \qquad (6.3.1)$$

and there are two equations of motion:

$$\frac{\partial L}{\partial \theta} - \frac{d}{dt}\left(\frac{\partial L}{\partial \dot{\theta}}\right) = 0 \qquad (6.3.2a)$$

and

$$\frac{\partial L}{\partial \phi} - \frac{d}{dt}\left(\frac{\partial L}{\partial \dot{\phi}}\right) = 0, \qquad (6.3.2b)$$

one for each degree of freedom. These, given the above Lagrangian (6.3.1), reduce to

$$\sin\theta\cos\theta\dot{\phi}^2 - \frac{g\sin\theta}{\ell} = \ddot{\theta} \qquad (6.3.3a)$$

and

$$0 = \frac{d}{dt}[\sin^2\theta\dot{\phi}], \qquad (6.3.3b)$$

the desired results.

Two first integrals follow readily from symmetries of the Lagrangian. Since ϕ is an ignorable coordinate, equation (6.3.3b) is integrated immediately to give

$$\sin^2\theta\dot{\phi} = C_1 \text{ (a constant).} \qquad (6.3.4)$$

Furthermore, since the Lagrangian is not a function of the independent variable t, the quantity

$$\dot{\theta}\frac{\partial L}{\partial \dot{\theta}} + \dot{\phi}\frac{\partial L}{\partial \dot{\phi}} - L = D \text{ (a constant)}, \qquad (6.3.5)$$

that is, given the particular Lagrangian (6.3.1),

$$\frac{m(\ell^2\dot{\theta}^2 + \ell^2\sin^2\theta\dot{\phi}^2)}{2} + mg\ell(1 - \cos\theta) = D, \qquad (6.3.6)$$

which is equivalent to energy conservation. In dynamical problems first integrals, like C_1 or D, are called *constants of motion*. Here, the constant C_1 is proportional to the vertical component of the angular momentum, and the constant D is equal to the energy. As such, they constitute a partial integration of the equations of motion. Problem 6.5 *Generalized Kepler's Third Law*, problem 6.6 *2-D Harmonic Motion*, and problem 6.7 *Elastic Pendulum* are related examples.

6.4 LAGRANGE AND HAMILTON

Working directly from the Principle of Least Action, Lagrange was the first to develop a general variational method for dynamics and the first to derive many of its consequences. Lagrange made his initial discoveries while an Italian youth of nineteen and twenty years of age in 1849–50. He sent his results to Euler, the recognized authority in the field and, in turn, "Euler sent [them] . . . to his official superior Maupertuis, then president of the Berlin Academy. Finding in Lagrange an unexpected defender of his Principle of Least Action, Maupertuis arranged for him to be offered, at the earliest opportunity, a chair of mathematics in Prussia, ... This proposition, transmitted through Euler, was rejected out of shyness; and nothing ever came of it."[3] But much came of Lagrange's discoveries. Maupertuis had dreamed and the great Euler labored, but Lagrange actually completed, with the publication of *Mecanique Analytic* in 1811, a dynamical theory based upon the Principle of Least Action.

How is Lagrange's lifetime work in dynamics [Lagrange multipliers (chap. 4), the Lagrangian function (chap. 6), a systematic approach to generalized coordinates (chap. 7), and many-particle systems (chap. 7)] best evaluated? No more apt an assessment can be given than that of Hamilton: "Lagrange has perhaps done more than any other analyst, to give extent and harmony to such deductive researches, by showing that the most varied consequences respecting the motions of systems of bodies may be derived from one radical formula; the beauty of the method so suiting the dignity of the results, as to make his great work a kind of scientific poem."[4] To go beyond Lagrange's "poem" one must

---

[3]*Dictionary of Scientific Biography*, Charles Coulton Gillespie, ed. (New York: Scribners, 1971), vol. 7, p. 561.

[4]"On a General Method in Dynamics," Philosophical Transactions of the Royal Society (1835), p. 95, quoted in Yourgrau and Mandelstam, *Variational Principles in Dynamics and Quantum Theory* (New York: Dover, 1968), p. 44.

broaden the physical base of the Principle of Least Action. Hamilton working in the 1830s did just that. Accordingly, we turn once more to Hamilton's Principle.

CHAPTER 6 PROBLEMS

Problem 6.1 *Redux*
Show how the "Snell's Law" of particle mechanics (6.1.8a) and energy conservation (6.1.8b) each follow, respectively, from the necessary conditions, (6.1.7a) and (6.1.7b), for making the relevant Hamilton's first principal function S stationary.

Problem 6.2 *Equations of Motion*
The context is section 6.2. From the three Euler-Lagrange equations (6.2.9a,b,c) and conservation of energy (6.2.6) derive the four equations $m\ddot{\mathbf{x}} = -\nabla\mathbf{U}$ and $\lambda = -1$.

Problem 6.3 *An Optical Lagrangian?*
The context is sections 5.4 and 6.2. Work out the optical analog of the procedure used to derive the single particle Lagrangian and Lagrange's equations of motion. Show that the Lagrange multiplier in the optical case vanishes.

Problem 6.4 *Other Coordinates*
Here we consider other coordinates q_1, q_2, and q_3, alternative to Cartesian coordinates x, y, and z, which also completely describe a particle's position. Suppose the two sets of coordinates are related by the time independent, reversible transformations:

$$x = x(q_1, q_2, q_3), \, y = y(q_1, q_2, q_3), \, z = z(q_1, q_2, q_3)$$

$$q_1 = q_1(x, y, z), \, q_2 = q_2(x, y, z), \, q_3 = q_3(x, y, z).$$

Spherical (r, θ, ϕ) and cylindrical (r, θ, z) coordinates are related to Cartesian coordinates by transformations of this kind.

(a) Show that the kinetic energy

$$T = \frac{m}{2}(\dot{x}^2 + \dot{y}^2 + \dot{z}^2)$$

transforms into a homogeneous quadratic function of the velocities $\dot{q}_i$;

of the form

$$T = \frac{m}{2}[a_{11}\dot{q}_1^2 + a_{22}\dot{q}_2^2 + a_{33}\dot{q}_3^2$$
$$+ (a_{12} + a_{21})\dot{q}_1\dot{q}_2 + (a_{13} + a_{31})\dot{q}_1\dot{q}_3 + (a_{32} + a_{23})\dot{q}_3\dot{q}_2],$$

i.e.,

$$T = \sum_{i,j}^{3,3}(a_{ij}\dot{q}_i\dot{q}_j)$$

where the "matrix elements" a_{ij} are given by

$$a_{ij} = \left(\frac{\partial x}{\partial q_i}\right)\left(\frac{\partial x}{\partial q_j}\right) + \left(\frac{\partial y}{\partial q_i}\right)\left(\frac{\partial y}{\partial q_j}\right) + \left(\frac{\partial z}{\partial q_i}\right)\left(\frac{\partial z}{\partial q_j}\right).$$

(b) Given that the kinetic energy is a homogeneous quadratic function of the velocities $\dot{q}_i$, show that

$$\sum_{k=1}^{3}\dot{q}_k\frac{\partial T}{\partial \dot{q}_k} - T = T$$

for T expressed in any set of coordinates q_k ($i = 1 \ldots 3$). From this we can conclude that the form of the augmented action found in equation (6.2.14), that is, $A^* = \int_{t_1}^{t_2}(T - U)\,dt$, follows from the procedure of section 6.2 when the Lagrangian $L = T - U$ is expressed in terms of any coordinate system which is related to Cartesian coordinates by a time-independent reversible transformation.

Problem 6.5 *Generalized Kepler's Third Law*
A particle of mass m is constrained to move in a plane under the influence of a central force derived from a potential $U(r)$ where r is the distance in the plane from a fixed force center at $r = 0$.
   (a) Write the Lagrangian in (r, θ) plane polar coordinates.
   (b) Show that the constant of the motion corresponding to the fact that θ is an ignorable coordinate is

$$mr^2\dot{\theta} = \ell \text{ (a constant)}$$

which in turn is equivalent to

$$\ell = 2m\frac{dA}{dt}$$

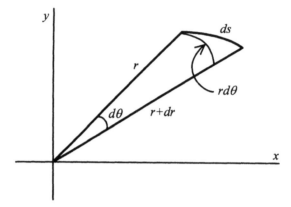

Fig. 6.1. Kepler's Third Law: $dA/dt = $ constant
where $dA = r^2 d\theta/2 + O(d\theta\, dr)$ so that $dA/dt = r^2\dot\theta/2$

where dA/dt is the rate at which the mass position vector r sweeps out area. See figure 6.1.

Problem 6.6 *2-D Harmonic Motion*

A mass m is attached to a point at the origin by a spring and required to move in a horizontal plane. Assume the relaxed length of the spring is ignorably small so that the potential energy stored is given to good approximation by $U(r) = kr^2/2$.

(a) Write the Lagrangian of this system in (r, θ) plane polar coordinates, from it determine Lagrange's equations of motion for $r(t)$ and $\theta(t)$, and reduce these to a single uncoupled, second order, differential equation for $r(t)$.

(b) Integrate this differential equation for $r(t)$ once. Show that this result is equivalent to the first integral following from the fact that L does not depend explicitly on the time.

(c) Show that the special orbits: a circle $[r = $ constant, $\theta \propto t]$ and straight-line motion through the origin $[\theta = $ constant, $r \propto \sin(\text{const.}t + \text{const.})]$, are possible solutions.

Problem 6.7 *Elastic Pendulum*

An elastic pendulum of spring constant k, relaxed length l_o, and bob mass m is constrained to oscillate in a vertical plane.

(a) Write the Lagrangian of the pendulum bob in plane polar coordinates (r, θ).

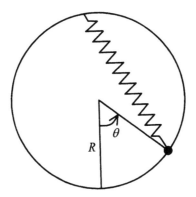

Fig. 6.2. Circle with mass

(b) Generate the equations of motion.

(c) Find the frequency of small oscillations in the r and θ coordinates around the position of stable equilibrium. [Answer: $\omega_r = \sqrt{k/m}$ and $\omega_\theta = \sqrt{g/(l_o + mg/k)}$].

Problem 6.8 *Circle*

A mass m is constrained to move on a circle of radius R and also is attached to the top of the circle with a spring of constant k and relaxed length l_o. Use the angle θ to indicate the particle position. See figure 6.2.

(a) Write the potential energy function $U(\theta)$ and the Lagrangian L.

(b) Under what conditions are there $\theta \neq 0$ equilibria?

(c) Find Lagrange's equation of motion.

(d) Find the frequency of small oscillations around one of the $\theta \neq 0$ equilibria.

Problem 6.9 *Ball within Circle*

A solid ball of uniform composition, mass m, and radius R is constrained to roll without slipping in a vertical plane on the inside of a circle of radius L. See figure 6.3 for definitions of coordinates θ and ϕ describing this rolling analog of a pendulum.

(a) Given that the sphere kinetic energy $T = m V_{cm^2}/2 + I_{cm}\omega^2/2$ where ω is the ball angular velocity and $I = (2/5)m R^2$ is its moment of inertia, write the Lagrangian in terms of the angle θ locating the position of the ball's center of mass.

(b) Find the Euler-Lagrange equation and the frequency of small oscillations around the equilibrium at $\theta = 0$.

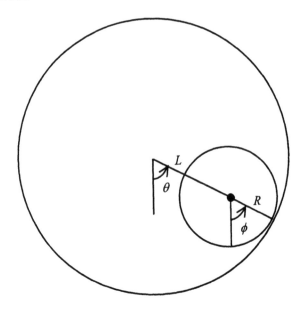

Fig. 6.3. Ball rolling within a circle

(c) If the period of small oscillation of a simple pendulum of length *L* and mass *m* is 1 sec, what is the period of a small oscillation around the equilibrium of this rolling ball?

Hamilton's Principle—Extended

> Accordingly, from this known law of least or
> stationary action, I deduced (long since) another
> connected and coextensive principle, which may
> be called, by analogy, the Law of Varying
> Action, and which seems to offer naturally a
> method such as we are seeking.
> —W. R. Hamilton, *Dublin University Review,*
>   1833

HAMILTON'S PRINCIPLE applies even to systems of particles which do not conserve energy. The formalism of multiparticle systems and generalized coordinates is developed.

7.1 HAMILTONIAN SYSTEMS

Sometimes a principle established for one set of circumstances applies more widely, or a technique effective in one context is effective in others as well. The history of the principle of relativity provides an example. Mechanical laws obey a relativity principle: if $F = ma$ is valid in one reference frame, it is also valid in all other frames translating uniformly with respect to the first. Yet nineteenth-century physicists believed that optical phenomena did not obey a relativity principle and that the laws of optics took their simplest form in a reference frame at rest with respect to an all-pervading ether. Then, in 1905, Albert Einstein introduced the special theory of relativity. This theory extended the principle of relativity to all physical laws, including optics, and had no use for the concept of ether.[1]

Again, the relation between electric field and charge distribution (Gauss's Law) was first suggested and then confirmed by experiments

---

[1] The conservative character of special relativity in extending an old and restricted principle to a wider class of phenomena is argued brilliantly by Hermann Bondi in his popularization *Relativity and Common Sense* (New York: Dover, 1964), chap. 1.

on stationary charges. Later, Gauss's Law was found to apply without approximation to charges in motion. In these ways old and limited laws extend their domain.

Such is the case with Hamilton's Principle. Recall that we derived the Restricted Hamilton's Principle from the Principle of Least Action and that the latter applies only when the total mechanical energy is a constant of the motion. However, Hamilton's Principle produces the correct equations of motion even on occasions when the energy of the particle or system of particles is not conserved yet still can be represented in terms of a potential function! Such occasions arise when forces represented by constraints do work on the system or when the potential energy is an explicit function of time.

In fact, Hamilton's Principle applies to all systems, called *Hamiltonian systems*, whose dynamics is fully governed by a potential energy function and constraints. A particle constrained by gravity and normal forces to remain on a surface is Hamiltonian because the influence of these two forces can be represented by a potential function and a constraint. An electron under the influence of a slowly oscillating electric field is also Hamiltonian because the force on the electron is the negative gradient of a time dependent potential function. On the other hand, a particle on which kinetic friction exerts a significant influence is not Hamiltonian because kinetic friction can be represented neither by a potential function nor by a constraint.

The extended Hamilton's Principle claims that a true trajectory makes Hamilton's first principal function S stationary. Recall that

$$S = \int_{t_1}^{t_2} L\,dt. \tag{7.1.1}$$

If the Lagrangian L is not a function of the independent integration variable t, so that $\partial L/\partial t = 0$, then the system is said to be *conservative* and

$$\sum_{i=1} \dot{q}_i \frac{\partial L}{\partial \dot{q}_i} - L = D, \tag{7.1.2}$$

a constant of the motion. Here the coordinates q_i specify the state of the system. Furthermore, if $\partial L/\partial t = 0$, if the kinetic energy T is a homogeneous quadratic function of the "velocities" $\dot{q}_i$, and if the potential energy U is a function of the coordinates q_i alone so that $U = U(q_1 \ldots)$, then L is said to be in *natural form* and energy conservation,

$$T + U = D \text{ (a constant)}, \tag{7.1.3}$$

follows immediately from (7.1.2).[2] Otherwise, the left-hand side of equation (7.1.2) need not be the total energy $T + U$; a system could be conservative yet not conserve energy.

Suppose, for example, that $L = T - U$ where $T = z^3 \dot{x}^2 + \dot{x} \dot{y}$ and $U = U(x, y, z)$. Then L represents a conservative system, is in natural form, and the energy $T + U = z^3 \dot{x}^2 + \dot{x} \dot{y} + U(x, y, z)$ is a constant of the motion. Problem 7.1 *Classification* contains other examples.

7.2 WATT'S GOVERNOR

Watt's Governor is a popular example of a one-particle, conservative, Hamiltonian system which does not conserve energy. It is essentially (see fig. 7.1) a circular wire frame which rotates with angular velocity ω around its vertical symmetry axis. A massive bob can slip without friction along the frame but is constrained to remain on it. This contraption is one of James Watt's (1736–1819) many improvements of the steam engine. Its purpose is to regulate the rotation speed of a steam engine driveshaft. If the steam powered shaft (lying along the symmetry axis) turns too quickly, the bob moves outward away from the rotation axis (toward $\theta = \pi/2$) and, as a result of linkages not shown, diminishes the steam supply to the engine. Here we ignore the basic regulating function of Watt's Governor, assume ω remains constant, and use Hamilton's Principle to determine the dynamics of the massive bob. See problems 7.2 *Undriven Watt's Governor* and 7.3 *Rotating Tube* for variations.

The dynamics of the driven Watt's Governor is completely determined by its gravitational potential energy function

$$U(\theta) = -mgR \cos \theta. \tag{7.2.1}$$

---

[2]A further distinction among kinetic energy, potential energy, and coordinate transformations which contain the time explicitly (*rheonomic*) and those which don't (*scleronomic*) is often made. But dynamical consequences follow only from the absence or presence of time dependence in the complete Lagrangian, that is, from whether or not the system is conservative or not, and not necessarily from the absence or presence of time in the Lagrangian's parts. For example, a system may be scleronomic in both the kinetic energy and the potential energy yet still have an explicitly time independent Lagrangian. Lanczos attributes the words scleronomic and rheonomic to Ludwig Boltzman (1844–1906). See *The Variational Principles of Mechanics* (New York: Dover, 1949), p. 32. Scleronomic (meaning "hardened") derives from the Greek *sclerosis* ("hardening"), as in arterial *sclerosis*. Rheonomic ("flowing") also comes from the Greek.

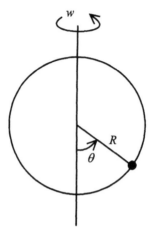

Fig. 7.1. Watt's governor

The position of the wire frame, measured by the angle $\phi(t)$, is constrained by

$$\phi(t) = \phi_o + \omega t. \tag{7.2.2}$$

Therefore, the bob's kinetic energy

$$T = \frac{m}{2}(R^2\dot\theta^2 + R^2\omega^2 \sin^2\theta), \tag{7.2.3}$$

and its Lagrangian

$$L(\theta, \dot\theta) = \frac{m}{2}(R^2\dot\theta^2 + R^2\omega^2 \sin^2\theta) + mgR\cos\theta. \tag{7.2.4}$$

We have a one-dimensional (θ), conservative (since $\partial L/\partial t = 0$), Hamiltonian system. However, the Lagrangian (7.2.4) is not in natural form since the kinetic energy (7.2.3) is not a homogeneous quadratic function of $\dot\theta$. Therefore, the quantity

$$\dot\theta\frac{\partial L}{\partial \dot\theta} - L = \frac{m}{2}(R^2\dot\theta^2 - R^2\sin^2\theta\omega^2) - mgR\cos = D \tag{7.2.5}$$

is a constant of the motion but is not the bob's energy E. Rather, D can be interpreted as the sum of the kinetic energy in a reference frame rotating with the wire frame $mR^2\dot\theta^2/2$, the gravitational potential energy $-mgR\cos\theta$, and a potential energy $-mR^2\omega^2\sin^2\theta/2$ associated with a noninertial "centrifugal force." The energy E in an inertial reference frame,

$$E = \frac{m}{2} R^2 \dot{\theta}^2 + \frac{m}{2} R^2 \omega^2 \sin^2 \theta - mgR \cos \theta, \qquad (7.2.6)$$

is not conserved. Rather, according to (7.2.5) and (7.2.6),

$$E = D + m R^2 \omega^2 \sin^2 \theta. \qquad (7.2.7)$$

As the bob moves away from the rotation axis, its speed increases because the wire frame does positive work on the bob. It is, in fact, common to say that the constraint does the work.

Lagrange's equation of motion,

$$\frac{\partial L}{\partial \theta} - \frac{d}{dt}\left(\frac{\partial L}{\partial \dot{\theta}}\right) = 0, \qquad (7.2.8)$$

for this system reduces to

$$mR^2 \ddot{\theta} = m R(R\omega^2 \cos \theta - g) \sin \theta, \qquad (7.2.9)$$

which can be interpreted as the component of $F = ma$ tangent to the circular wire frame. Because this second-order, ordinary, differential equation is nonlinear in θ, we'll be satisfied with identifying its equilibria and determining their stability properties.

The equilibria θ_{eq} are angles at which the right-hand side of the equation of motion (7.2.9) vanishes. In the domain $0 \leq \theta \leq \pi$

$$\theta_{eq} = 0, \pi, \text{ and } \cos^{-1}\left(\frac{g}{R\omega^2}\right). \qquad (7.2.10)$$

The third equilibrium, $\theta_{eq} = \cos^{-1}(\frac{g}{R\omega^2})$, is a dynamic one associated with motion at constant θ and is a possibility only when $g \leq R\omega^2$. We suspect that the second equilibrium, $\theta_{eq} = \pi$, is unstable.

In calculating the stability properities of each of these equilibria, we expand the equation of motion (7.2.9) around $\theta = \theta_{eq}$ through first order in the deviation $\theta - \theta_{eq}$. Doing so, we find that

$$mR^2 \ddot{\theta} = -mR[R\omega^2(1 - 2\cos^2 \theta_{eq}) + g \cos \theta_{eq}](\theta - \theta_{eq}). \qquad (7.2.11)$$

When $\theta = \pi$, equation (7.2.11) becomes

$$mR^2 \ddot{\theta} = mR[R\omega^2 + g](\theta - \theta_{eq}) \qquad (7.2.12)$$

99

whose solution is

$$(\theta - \pi) \propto \exp(\pm \gamma t) \qquad (7.2.13)$$

where the exponential growth rate

$$\gamma = \sqrt{\omega^2 + g/R}. \qquad (7.2.14)$$

Therefore, we expect any small deviation from the equilibrium $\theta_{eq} = \pi$ to grow exponentially. Therefore, this solution is unstable. The equation of motion (7.2.11) expanded around $\theta_{eq} = 0$ is

$$m R^2 \ddot{\theta} = -m R[-R\omega^2 + g]\theta, \qquad (7.2.15)$$

which also has an unstable solution with growth rate

$$\gamma = \sqrt{\omega^2 - g/R} \qquad (7.2.16)$$

when $g/R < \omega^2$. On the other hand, if $g/R > \omega^2$, solutions to (7.2.15),

$$\theta \propto \exp(\pm i \varpi t), \qquad (7.2.17)$$

are periodic with angular oscillation frequency

$$\varpi = \sqrt{g/R - \omega^2}. \qquad (7.2.18)$$

Any small deviation from the equilibrium $\theta_{eq} = 0$ whenever $g/R > \omega^2$ would lead to oscillations around $\theta_{eq} = 0$ with angular frequency $\varpi = \sqrt{g/R - \omega^2}$. The same analysis tells us that the $\theta_{eq} = \cos^{-1}(\frac{g}{R\omega^2})$ equilibrium (possible only when $g < R\omega^2$) is stable with oscillation frequency

$$\varpi = \omega \sqrt{1 - \left(\frac{g}{R\omega^2}\right)^2}. \qquad (7.2.19)$$

The stability properties of the driven Watt's Governor are summarized in figure 7.2.

7.3 MULTIPARTICLE SYSTEMS

Physics is the science of interacting particles and only in a special circumstance can we focus on one particle to the exclusion of others: when the mass m of one particle in an interacting pair is much less than the mass

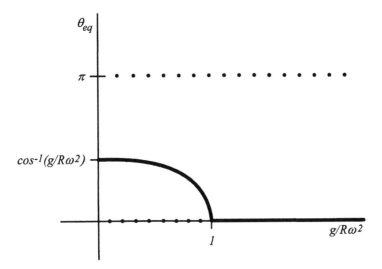

Fig. 7.2. Stability of the driven Watt's governor: the solid line shows stable equilibria; the filled circles unstable equilibria. (After fig. 4.5-2 in *Newtonian Dynamics* by Ralph Baierlein [New York: McGraw-Hill, 1983], p. 134.)

M of the other particle, so that $m \ll M$. Then the influence of the heavier on the lighter particle is approximated by a potential energy function *U* of the coordinates of the lighter particle alone. This approximation is behind analyses of single-particle dynamics. But single-particle dynamics is not generally adequate. Often the potential is a sensitive function of the coordinates of both or of several interacting particles. Consider two particles composing a diatomic molecule, a large number of molecules bound together in a crystal, or a planetary system of several planets and a star. The coordinates describing a multiparticle system can also be linked by constraints characteristic of rigid bodies, connecting ropes, or restraining surfaces. In these instances a system is formed and a multiparticle version of Hamilton's Principle is required.

The Lagrangian of a single particle is the difference between the particle's kinetic *T* and potential *U* energies. Likewise, the Lagrangian of an *N*-particle system is the difference between the total kinetic energy of the system and its total potential energy,

$$L = \sum_{j=1}^{N} \frac{m_j v_j^2}{2} - U(q_1, q_2, \ldots q_{3N}), \qquad (7.3.1)$$

where the index j denotes particle number, $j = 1 \ldots N$. In the absence of constraints, $3N$ coordinates are needed to completely specify an N–particle system, and the potential U is generally a function of all of these. While the form of (7.3.1) is plausible, it also follows from a multiparticle action,

$$A = \sum_{i=1}^{N} \int m_i v_i \, ds_i, \qquad (7.3.2)$$

given the constraint of energy conservation. The derivation of (7.3.1) from (7.3.2) follows the pattern of section 6.2. See problem 7.4 *Multiparticle Lagrangian*. But beware. Hamilton's Principle *per se* cannot be derived from the Principle of Least Action. Any derivation of (7.3.1) from (7.3.2) makes sense only when energy is conserved yet Hamilton's Principle applies even when constraints do work and energy is not conserved. Therefore, such derivations are only heuristic, that is, suggestive or helpful in discovery, rather than logically compelling.

Truly fundamental principles cannot, of course, be derived from anything more fundamental. Rather, they are underived givens with consequences much as theorems are consequences of the postulates in a deductive mathematical system. Although fundamental principles are not derived, in science they can and should be verified either directly or indirectly through careful experimentation and observation. In just this sense is Hamilton's Principle fundamental; it has important consequences: $F = ma$ when the forces are derivable from a potential, the Principle of Least Action, and the Principle of Least Potential Energy, all of which have been well verified.

Variational and Newtonian mechanics not only have different foundations (Hamilton's Principle versus Newton's Laws of Motion), the history of their development is also different. Newtonian mechanics is grounded on $F = ma$. The latter remained unmodified for two hundred years while various applications were worked out in detail. In contrast, variational mechanics was reinvented several times (first as Maupertuis's Principle of Least Action, then as Lagrangian mechanics, then . . .) during the eighteenth century. With the discovery of Hamilton's Principle in 1833 the stage was set for exploring its many consequences. We are at that stage now.

7.4 TWO-BODY CENTRAL POTENTIAL

Two particles interact via a central potential $U(r)$ that depends only upon the particle separation r. Universal gravitation and Coulomb potentials take this form. Since each particle has three degrees of freedom, the two-particle system has six degrees of freedom. We suspect the Lagrangian will be most simple when the particle positions are expressed in terms of so-called *center of mass coordinates* denoting the system's center of mass, (X, Y, Z), and separation, r, and orientation, θ and ϕ, of a line joining the two particles. The directions corresponding to zeros of θ and ϕ are arbitrary. We'll take advantage of this fact soon.

The center of mass and separation vectors, $\mathbf{R}$ and $\mathbf{r}$, respectively, are related to Cartesian coordinates, $\mathbf{x}_1$ and $\mathbf{x}_2$, by

$$\mathbf{R} = \frac{m_1\mathbf{x}_1 + m_2\mathbf{x}_2}{M} \tag{7.4.1a}$$

and

$$\mathbf{r} = \mathbf{x}_2 - \mathbf{x}_1 \tag{7.4.1b}$$

where $M = m_1 + m_2$ is the system's total mass. Inverting these equations we get

$$\mathbf{x}_1 = \mathbf{R} - \frac{m_2}{M}\mathbf{r} \tag{7.4.2a}$$

and

$$\mathbf{x}_2 = \mathbf{R} + \frac{m_1}{M}\mathbf{r}. \tag{7.4.2b}$$

The system's kinetic energy is, therefore, given by

$$T = \frac{m_1\dot{x}_1^2}{2} + \frac{m_2\dot{x}_2^2}{2} \tag{7.4.3}$$

which expressed in terms of coordinates, $(X, Y, Z, r, \theta, \phi)$, is

$$T = \frac{M(\dot{X}^2 + \dot{Y}^2 + \dot{Z}^2)}{2} + \frac{\mu}{2}\left(\frac{ds}{dt}\right)^2. \tag{7.4.4}$$

Here the *reduced mass*

$$\mu = \frac{m_1 m_2}{m_1 + m_2} \tag{7.4.5}$$

and

$$ds^2 = d\mathbf{r} \cdot d\mathbf{r} = dr^2 + r^2 d\theta^2 + r^2 \sin^2\theta d\phi^2. \tag{7.4.6}$$

Because X, Y, and Z are ignorable coordinates of the Lagrangian L where

$$L = \frac{M(\dot{X}^2 + \dot{Y}^2 + \dot{Z}^2)}{2} + \frac{\mu}{2}(\dot{r}^2 + r^2\dot{\theta}^2 + r^2\sin^2\theta\dot{\phi}^2) - U(r), \quad (7.4.7)$$

the derivatives $\partial L/\partial\dot{X}$, $\partial L/\partial\dot{Y}$, and $\partial L/\partial\dot{Z}$, that is, the components of the total linear momentum, $M\dot{X}$, are constants. Therefore, a reference frame with origin at the center of mass (in which the system momentum vanishes) is also an inertial reference frame. Adopting this frame, the Lagrangian (7.4.7) reduces to

$$L = \frac{\mu}{2}(\dot{r}^2 + r^2\dot{\theta}^2 + r^2\sin^2\theta\dot{\phi}^2) - U(r) \quad (7.4.8)$$

which can be interpreted as that of a single particle with reduced mass μ.

Lagrange's equations of motion, one for each dependent variable, $r(t)$, $\theta(t)$, and $\phi(t)$, are, respectively,

$$\mu\ddot{r} = -\frac{\partial U}{\partial r} + \mu r\dot{\phi}^2\sin^2\theta, \quad (7.4.9a)$$

$$\frac{d}{dt}[\mu r^2\dot{\theta}] = \mu r^2\dot{\phi}^2\sin\theta\cos\theta, \quad (7.4.9b)$$

and

$$\frac{d}{dt}[\mu r^2\dot{\phi}\sin^2\theta] = 0. \quad (7.4.9c)$$

The last, (7.4.9c), integrates immediately to

$$\mu r^2\dot{\phi}\sin^2\theta = C_1. \quad (7.4.10)$$

Since the Lagrangian (7.4.8) is not an explicit function of the time, the kinetic energy T is a homogeneous quadratic function of the velocities, and the potential is a function of only the coordinates, the Lagrangian is in natural form, and the energy E is a constant of the motion given by $E = \sum\dot{q}_i(\partial L/\partial\dot{q}_i) - L$, that is, by

$$E = \frac{\mu(\dot{r}^2 + r^2\dot{\theta}^2 + r^2\sin^2\theta\dot{\phi}^2)}{2} + U(r). \quad (7.4.11)$$

Each of the three equations of motion (7.4.9a,b,c) are second-order differential equations and their complete solution requires six constants of motion. In addition to (7.4.10) and (7.4.11), another constant of motion is given by

$$r^4[\dot{\theta}^2 + \dot{\phi}^2 \sin^2 \theta] = C_2 \qquad (7.4.12)$$

which is straightforwardly verified by taking the time derivative of (7.4.12) and using (7.4.9a,b,c). We display (7.4.12) in order to emphasize that if we choose (as we are free to do) the polar axis of our coordinate system so that $\theta = 90°$ at the moment when $\dot{\theta} = 0$, then the constant in (7.4.12) is zero and $\theta = 90°$ for all time. This conclusion also follows from a direct inspection of the equation of motion (7.4.9b). If we choose $\theta = 90°$ when $\dot{\theta} = 0$ in (7.4.9b), then $\ddot{\theta} = 0$ at this instant. Therefore, $\theta = 90°$ always and the motion of the two-particle system interacting through a central potential is confined to a plane. Given $\dot{\theta} = 0$ and $\theta = 90°$, the three equations (7.4.9a,b) and (7.4.10) reduce to two,

$$\mu\ddot{r} = -\frac{\partial U}{\partial r} + \mu r\dot{\phi}^2 \qquad (7.4.13)$$

and

$$\mu r^2\dot{\phi} = \ell \text{ (a constant)}. \qquad (7.4.14)$$

Our notation recognizes that the constant $\mu r^2\dot{\phi}$ is the system angular momentum ℓ reckoned from the center of mass. Using (7.4.14) to eliminate $\dot{\phi}$ from the equation of motion (7.4.13) gives

$$\mu\ddot{r} = -\frac{\partial U}{\partial r} + \frac{L^2}{\mu r^3} \qquad (7.4.15)$$

or, equivalently,

$$\mu\ddot{r} = -\frac{\partial}{\partial r}\left[U + \frac{\ell^2}{2\mu r^2}\right]. \qquad (7.4.16)$$

The quantity in the brackets of (7.4.16),

$$U_{eff}(r) = U(r) + \frac{\ell^2}{2\mu r^2}, \qquad (7.4.17)$$

is an *effective potential*. Solutions $r(t)$ and $\phi(t)$ of equations (7.4.14) and (7.4.16) describe the dynamics of a two-particle system with potential $U(r)$.

If only the shape of the orbit is desired, one can eliminate the time t from among the three variables r, ϕ, and t in the solutions $r(t)$ and $\phi(t)$. In practice, this is done most readily by using the identity

$$\dot{r}(t) = \dot{\phi}(t)\frac{dr}{d\phi} \qquad (7.4.18)$$

to eliminate time derivatives from either the equation of motion, (7.4.16), or from energy and angular momentum conservation, (7.4.11) and (7.4.14), with $\theta = 90°$. Alternatively, the orbit shape is determined directly from the Principle of Least Action. See problem 5.4 *Orbit Shapes*. Problem 7.6 *Compound Pendulum* and problem 7.7 *3-D, 2-Mass Harmonic Oscillator* also involve multiparticle systems.

7.5 GENERALIZED COORDINATES

In order that Lagrange's equations of motion be deduced from Hamilton's Principle, the dependent variables $q_i(t)$, $i = 1 \ldots n$, must be independent of one another. If instead the system's n coordinates are related through a constraint equation

$$C(q_1 \ldots q_n) = \text{constant}, \qquad (7.5.1)$$

the deduction fails. This is so because in varying Hamilton's first principal function we arrive, according to the general method outlined in section 2.4, at the condition

$$0 = \int_{t_1}^{t_2} \sum_{i=1}^{n} \eta_i(t) \left[\left(\frac{\partial L}{\partial q_i} \right) - \frac{d}{dt} \left(\frac{\partial L}{\partial \dot{q}_i} \right) \right] dt. \qquad (7.5.2)$$

Because the constraint (7.5.1) connects the coordinates $q_i(t)$, it also connects the comparison and difference functions $Q_i(t)$ and $\eta_i(t)$. But if and only if each of the $\eta_i(t)$, $i = 1 \ldots n$, is an independent arbitrary function, do n Lagrange equations of motion,

$$\left(\frac{\partial L}{\partial q_i} \right) - \frac{d}{dt} \left(\frac{\partial L}{\partial \dot{q}_i} \right) = 0, \qquad (7.5.3)$$

follow from (7.5.2) via the fundamental lemma. Constraints, like (7.5.1), that relate configuration space coordinates are *holonomic constraints*.[3] When holonomic constraints obtain, the full complement of Lagrange's equations do not follow from Hamilton's Principle and the fundamental

---

[3]*Nonholonomic* constraints cannot be so written. Examples of nonholonomic constraints are constraints which must be written as inequalities or as nonintegrable equations relating differentials of coordinates. The condition that a ball rolls without slipping is a nonholonomic constraint. See Lanczos, pp. 24 ff.

lemma. How then can constraints be incorporated into the variational process?

There are two distinct answers to this question. Let's consider an N-particle system with $n = 3N$ coordinates and m holonomic constraint equations,

$$C_j(q_1 \ldots q_n) = B_j \text{ (a constant)} \qquad (7.5.4)$$

where $j = 1 \ldots m$. First, we could solve the m constraints for m of the coordinates in terms of the others so that

$$q_j = q_j(q_1 \ldots q_{n-m}) \qquad (7.5.5)$$

where $j = 1 \ldots m$. Then, using equations (7.5.5), we eliminate the co-ordinates q_j from the Lagrangian leaving $n - m$ so-called *generalized coordinates*, $q_1 \ldots q_{n-m}$. Since generalized coordinates are, by construction, independent of one another, $n - m$ Lagrange equations of motion follow. A set of generalized coordinates is the smallest set of coordinates necessary to completely specify the state of a system. They are, in a descriptive sense, both concise and complete. While the number of generalized coordinates is unique for a particular system and equal to its degrees of freedom, the coordinates themselves are not unique. For instance, both the Cartesian coordinates, x and y, and the polar coordinates, r and θ, are generalized coordinates of a single particle constrained to a plane. The freedom to choose generalized coordinates most appropriate to the problem is a powerful tool of variational dynamics. Generalized coordinates are usually, but not always, the preferred approach.

A second approach is to retain the n non-generalized coordinates and incorporate the m homonomic constraints (7.5.4) with m Lagrange multipliers λ_j, $j = 1 \ldots m$. We might, for instance, see no reason to choose, out of n coordinates which specify a system with m constraints, a particular set of $n - m$ generalized coordinates. See, for instance, problem 7.8 *Ellipsoid of Revolution*. Then the integrand of the resulting augmented Hamilton's first principle function S^* is an augmented Lagrangian

$$\begin{aligned} L^*(q_1 \ldots q_n, q_1 \ldots q_n) &= L(q_1 \ldots q_n, q_1 \ldots q_n) \\ &+ \sum_{j=1}^{j=m} \lambda_j C_j(q_1 \ldots q_n). \end{aligned} \qquad (7.5.6)$$

The n coordinates are treated as independent variables. Then n

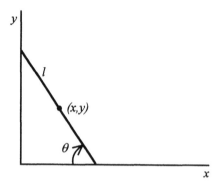

Fig. 7.3. Sliding ladder

augmented Lagrange's equations of motion

$$\left(\frac{\partial L}{\partial q_i}\right) - \frac{d}{dt}\left(\frac{\partial L}{\partial \dot{q}_i}\right) + \sum_{j=1}^{m} \lambda_j \frac{\partial C_j}{\partial q_i} = 0 \qquad (7.5.7)$$

follow where $i = 1 \ldots n,$. Finally, the m Lagrange multipliers λ_j are eliminated from the n equations (7.5.7) with the m constraint equations (7.5.4). We use both approaches in the next section.

7.6 SLIDING LADDER

A ladder of length ℓ and mass m, constrained to be in frictionless contact with two plane surfaces normal to one another, as shown in figure 7.3, will slide down until it strikes the bottom surface. The problem of deriving the ladder's equation of motion illustrates the difference between working with generalized coordinates and with Lagrange multipliers.

The ladder's kinetic energy is the sum of its kinetic energy under the assumption that the ladder mass m is concentrated at its center of mass (x, y) and the kinetic energy of rotation around an axis through the center of mass and normal to the plane of rotation. Thus the Lagrangian

$$L = \frac{m(\dot{x}^2 + \dot{y}^2)}{2} + \frac{I\dot{\theta}^2}{2} - mgy \qquad (7.6.1)$$

where I is the appropriate moment of inertia. The three coordinates x,

y, and θ are related by two holonomic constraints

$$x - \frac{\ell \cos \theta}{2} = 0 \qquad (7.6.2a)$$

and

$$y - \frac{\ell \sin \theta}{2} = 0. \qquad (7.6.2b)$$

Therefore, there is one degree of freedom and one generalized coordinate. If we choose the generalized coordinate to be θ and use (7.6.2a,b) to eliminate $\dot{x}$ and $\dot{y}$, then

$$L = \frac{m\ell^2 \dot{\theta}^2}{4} + \frac{I\dot{\theta}^2}{2} - \frac{mg\ell \sin \theta}{2}. \qquad (7.6.3)$$

Since L is not an explicit function of the independent variable t and is in natural form (that is, the kinetic energy is a homogeneous quadratic function of the generalized velocity $\dot{\theta}$ and the potential energy is a function of only the coordinates), the energy,

$$E = \frac{m\ell^2 \dot{\theta}^2}{4} + \frac{I\dot{\theta}^2}{2} + \frac{mg\ell \sin \theta}{2}, \qquad (7.6.4)$$

is a constant of the motion. The Euler-Lagrange equation following from (7.6.3) is

$$\left(\frac{m\ell^2}{2} + I \right) \ddot{\theta} = \frac{mg\ell \cos \theta}{2}, \qquad (7.6.5)$$

that is,

$$\ddot{\theta} = \frac{g/\ell}{1 + \frac{2I}{m\ell^2}} \cos \theta. \qquad (7.6.6)$$

Alternatively, we may retain all three coordinates x, y, and θ in the Lagrangian (7.6.1) and incorporate the two constraints (7.62a,b) into the variational process with two Lagrange multipliers, λ_x and λ_y. Making explicit use of the augmented Lagrange equations (7.5.7), we find

$$- m\ddot{x} + \lambda_x = 0, \qquad (7.6.7a)$$

$$- m\ddot{y} + \lambda_y - mg = 0, \qquad (7.6.7b)$$

and

$$- I\ddot{\theta} + \lambda_x \left(\frac{\ell \sin \theta}{2} \right) - \lambda_y \left(\frac{\ell \cos \theta}{2} \right) = 0. \qquad (7.6.7c)$$

The equations (7.6.7a,b,c) and constraints (7.6.2a,b) together are five equations containing five unknowns: the coordinates x, y, and θ and the Lagrange multipliers λ_x and λ_y. Let's eliminate four of the unknowns, x, y, λ_x, and λ_y, in favor of the fifth, θ. To do so, we take two time derivatives of the constraint equations (7.6.2a,b) and find

$$\ddot{x} = -\frac{\ell}{2}[\cos\theta\dot{\theta}^2 + \sin\theta\ddot{\theta}], \tag{7.6.8a}$$

and

$$\ddot{y} = \frac{\ell}{2}[-\sin\theta\dot{\theta}^2 + \cos\theta\ddot{\theta}]. \tag{7.6.8b}$$

When these are combined with (7.6.7a,b) we get

$$\lambda_x = -\frac{m\ell}{2}[\cos\theta\dot{\theta}^2 + \sin\theta\ddot{\theta}], \tag{7.6.9a}$$

and

$$\lambda_y = mg + \frac{m\ell}{2}[-\sin\theta\dot{\theta}^2 + \cos\theta\ddot{\theta}]. \tag{7.6.9b}$$

Finally, when combined with (7.6.7c) these yield the same equation of motion for θ, (7.6.6), as before.

7.7 SUMMARY AND PROSPECT

We have encountered Hamilton's Principle—a foundational principle of classical mechanics. But, more important, we are armed with a powerful method. Variational methods solve problems in all fields of physics. Their advantages are several: (1) they reduce the problem of generating governing differential equations to one of taking derivatives of a scalar function, (2) they allow us to use that particular set of coordinates best suited to the problem at hand, and (3) they easily incorporate the effects of constraints. In *Perfect Form* we have applied these methods to a number of problems in optics and mechanics.

The "variationalizing" of other branches of physics is an exciting prospect. To do so, we must consider what happens if the integral to be made stationary has more than one independent integration variable, that is, is a multiple integral. The unknown function might be a field quantity, say a time-independent wave function $\varphi(x, y, z)$, with the integration over each of the independent variables. The methods of chapter 2, slightly generalized, are sufficient for solving this problem. One finds that the resulting Euler-Lagrange equation is a partial differential equation with several independent variables—one for each integration variable x, y, and z. Given appropriate variational principles each with an associated multiple integral and scalar integrand, we can produce all the important partial differential equations of physics: the wave equation, the diffusion equation, Poisson's equation, Schrodinger's equation, and each of Maxwell's equations.

Finally, variational principles and methods invite us to theorize. Is there a phenomenon you wish to model? Does it conserve certain quantities? Try constructing the simplest scalar integrand which, used as a Lagrangian, results in the conservation of those quanities. Do you wish to consistently incorporate a new effect into an existing set of equations? Try doing so with Lagrange multipliers. What about unifying two theories? Maybe adding Lagrangians will work. Such thinking does bear fruit. General relativity and quantum mechanics both originated from variational principles.

CHAPTER 7 PROBLEMS

Problem 7.1 *Classification*
Identify which of the following kinetic and potential energies, T and U, result in a non-conservative system, which result in a conservative system with a first integral of the form $\sum_{i=1} \dot{q}_i \frac{\partial L}{\partial \dot{q}_i} - L = D$ (a constant) and which have natural forms in that the above conserved quantity reduces to the energy $T + U = E$ (a constant).

(a) $T = a(\dot{x}^2 + \dot{y}^2) + b\dot{x}\dot{y}$; $U = cxy + d\dot{x}$
(b) $T = a(\dot{x}^2 + \dot{x}\dot{y}) + bt$; $U = xy + bt$
(c) $T = ar^2\dot{\theta}^2$; $U = b\theta t^2$

Problem 7.2 *Undriven Watt's Governor*
The context is section 7.2. Allow the angle ϕ which indicates the angular position of the governor to be an unconstrained coordinate of the system along with θ which indicates the position of the massive bob. The wire frame is no longer massless but has a moment of inertia around the vertical symmetry axis of I so that its instantaneous kinetic energy is $I\omega^2/2$. The governor is initialized with an angular rotation frequency ω_0.

(a) Write the Lagrangian of this system as a function $L(\theta, \dot{\theta}, \dot{\phi})$,

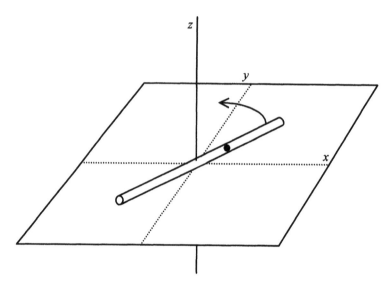

Fig. 7.4. Rotating tube

(b) determine the constants of the motion corresponding to the fact that $\partial L/\partial t = 0$ and to the ignorable coordinate ϕ,

(c) write the two equations of motion, and

(d) reduce these two to one by eliminating $\dot{\phi}$.

(e) Show that $\theta_o = \cos^{-1}(g/R\omega_o^2)$ is always a stable equilibrium around which the bob will oscillate with frequency $\varpi = \omega_o\sqrt{1 + (\frac{g}{R\omega_o^2})^2}$.

Here ω_o is the initial value of the angular frequency $\dot{\theta}$.

Problem 7.3 *Rotating Tube*

A particle of mass m is constrained to move inside a thin hollow frictionless tube (fig. 7.4) which is rotating with constant angular velocity ω in the horizontal xy plane about a fixed vertical axis normal to and through the center of the tube.

(a) Write the particle Lagrangian in plane polar coordinates.

(b) Identify any constants of the motion which arise from the form of the particle Lagrangian. Is energy conserved?

(c) Solve the relevant Euler-Lagrange equation and describe the particle's motion.

Problem 7.4 *Multiparticle Lagrangian*

The context is section 7.3. Derive the form of a multiparticle Lagrangian (7.3.1) by making stationary the multiparticle action (7.3.2), given that

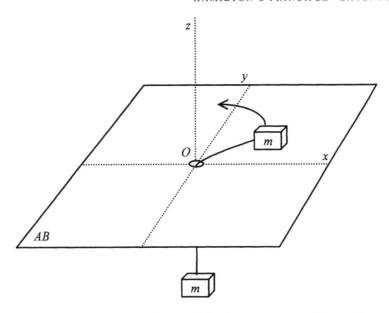

Fig. 7.5. The lower mass hangs while the upper mass slides without friction on the surface AB

the total energy is conserved. Follow the, appropriately generalized, procedure of section 6.2.

Problem 7.5 *Rotating Mass on a String*

In figure 7.5, AB represents a horizontal plane having a small opening at O. A string of length ℓ connects two particles each of mass m; one particle hangs freely from one end of the string while the other slides without friction on the surface. The particle on the surface is initially sliding in a circle of radius a with speed v_o. Let r and θ be the instantaneous plane polar coordinates describing the position of the sliding mass on the surface with respect to an origin at O. Find this system's two equations of motion.

Problem 7.6 *Compound Pendulum*

A compound pendulum (see fig. 7.6) is constrained to oscillate in a plane. Use the angles θ and ϕ for generalized coordinates. Find the pendulum's two equations of motion.

Problem 7.7 *3-D, 2-Mass Harmonic Oscillator*

Consider two masses, m_1 and m_2, connected by a spring with constant k and relaxed length l_o.

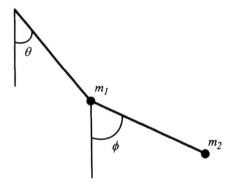

Fig. 7.6. Compound pendulum

(a) Write the Lagrangian in terms of center of mass Cartesian coordinates (X, Y, Z) and (r, θ, ϕ).

(b) Identify ignorable coordinates and the corresponding conserved quantities, and

(c) generate the equations of motion.

Problem 7.8 *Ellipsoid of Revolution*
A mass m is constrained to remain on the surface of an ellipsoid of revolution

$$a(x^2 + y^2) + bz^2 = 1$$

and is under the influence of a uniform gravitational force described by the potential $U = mgz$. Use the Principle of Least Potential Energy to find its equations of motion by writing the Lagrangian in terms of the rectangular coordinates (x, y, z) and introducing the above constraint with a Lagrange multiplier.

Index

Milton Keynes UK
Ingram Content Group UK Ltd.
UKHW020753150824
446937UK00006B/203